Plasmonics: Advanced Topics and Applications

Plasmonics: Advanced Topics and Applications

Edited by **Jonah Holmes**

New York

Published by NY Research Press,
23 West, 55th Street, Suite 816,
New York, NY 10019, USA
www.nyresearchpress.com

Plasmonics: Advanced Topics and Applications
Edited by Jonah Holmes

© 2015 NY Research Press

International Standard Book Number: 978-1-63238-361-7 (Hardback)

Contents

Preface

This book was inspired by the evolution of our times; to answer the curiosity of inquisitive minds. Many developments have occurred across the globe in the recent past which has transformed the progress in the field.

This book elucidates the theory as well as the practical applications of plasmonics and serves as a compilation of the modern advancements and researches in the field. It also covers the technical issues related to the field. The information presented in this book covers interesting topics of "modeling and computational techniques, focusing, guiding," and "plasmonic structures for light transmission". It will serve as a reference book for students, researchers and scientists engaged in the field.

This book was developed from a mere concept to drafts to chapters and finally compiled together as a complete text to benefit the readers across all nations. To ensure the quality of the content we instilled two significant steps in our procedure. The first was to appoint an editorial team that would verify the data and statistics provided in the book and also select the most appropriate and valuable contributions from the plentiful contributions we received from authors worldwide. The next step was to appoint an expert of the topic as the Editor-in-Chief, who would head the project and finally make the necessary amendments and modifications to make the text reader-friendly. I was then commissioned to examine all the material to present the topics in the most comprehensible and productive format.

I would like to take this opportunity to thank all the contributing authors who were supportive enough to contribute their time and knowledge to this project. I also wish to convey my regards to my family who have been extremely supportive during the entire project.

Editor

Modeling and Computational Methods for Plasmonics

Computational Electromagnetics in Plasmonics

Guy A. E. Vandenbosch

Additional information is available at the end of the chapter

1. Introduction

In electromagnetics, numerical techniques have been essential in the development of new technology in the last two decades. The rapidly growing computer capacity and calculation speeds make accurate solutions of very complex problems feasible. This has been especially true in the design of microwave and millimeter wave components and antennas. Whereas 30 years ago, the design of an antenna was based on simple analytical models, or trial and error strategies, nowadays, simulations seem to be as crucial to the design as real measurements.

The situation is quite different in plasmonics. Plasmonics is a quite novel research field and the application of computational electromagnetics in plasmonics can be categorized as "very recent". There are many challenges that still need to be faced and "missing links" that have to be solved.

The plasmonic structures targeted are structures in the order of magnitude of a wavelength at plasmonic frequencies, i.e. at near IR and optical frequencies, and beyond. Although this frequency range is totally different from the traditional range where computational tools have been developed, i.e. the microwave range, in most cases no special modeling techniques have to be used. By far most plasmonic topologies reported in literature have been analyzed / designed with the well-known numerical techniques implemented within in-house developed or commercial software packages. This means that in this chapter the major numerical techniques can be overviewed in a general sense, referring to standard literature. These techniques will not be derived or explained here in full detail. Instead, this chapter focuses on those aspects that come into the picture when the structure is plasmonic.

After the section on techniques and tools, this chapter will focus on the performance of these techniques and tools for plasmonic structures. This is done through an overview of benchmarks available in literature and by considering a few thoroughly analyzed structures. Missing links will be pointed out and suggestions will be given for the future.

2. Fundamental physical modeling differences

This section discusses the differences between classical topologies handled with computational electromagnetics and plasmonic topologies. Classical means structures at much lower frequencies, for example at microwave frequencies, where computational electromagnetics is a well-developed mature field.

2.1. Frequency

First of all, it is essential to point out that the interaction between light and plasmonic structures in the frequency bands considered can still be analyzed with a high degree of accuracy using classical electromagnetic theory. Although the frequency is orders of magnitude higher in plasmonics compared to microwaves, Mawell's laws are the same. They are linear, and thus scalable. As long as quantum effects do not have to be taken into account, which is still the case for the plasmonic applications considered, since the structures are not that small [1], there is no fundamental problem. The underlying formulation of Maxwell's equations can remain unaffected.

The fact that at this small scale, no quantum effects have to be taken into account is really a crucial observation. It means that the concept of a "scatterer", a device able to be excited by electromagnetic waves rather than particles, still works. Basically, the coupling between an electromagnetic (light-) wave and a plasmonic scatterer is thus the same as it is at microwave frequencies, and can be studied in the same way.

2.2. Volumetric currents

A consequence of the high frequencies is the small scale: the elementary building blocks of the topologies considered are at nanoscale. There are two important issues related to this. First, nanoscale fabrication technology of today is only able to generate 3D type structures (i.e. volumes). Thin 2D sheets, where the thickness of the metal is orders of magnitude smaller than the transversal dimensions of the pattern, as commonly used at microwave frequencies (for example a conducting strip or patch), are not possible. Second, for nanostructures operating at plasmonic frequencies the skin depth may be comparable with the structural dimensions, so that currents do flow over the complete volume and the device has to be described with volumetric currents. A surface current description is not sufficient.

2.3. Material characteristics

At microwave frequencies, in most cases the material properties are constant over the frequency bands considered. Also, apart from the losses, most metals behave more or less in the same way, i.e. as good conductors. At IR and optical frequencies however, most metals have very dispersive properties. Permittivities and conductivities may vary orders of magnitude over these frequency bands. The real part of the permittivity may even be negative. In Fig. 1, the permittivity of Cu and Ag are depicted, both real and imaginary part, which corresponds to conductivity. These data were obtained through experimental ellipsometry. Note that the difference between the two metals is enormous. It is evident that this strong variation has a serious impact on the behavior of a device over the frequency band. The variability of material characteristics thus has to be taken into account fully into the modeling.

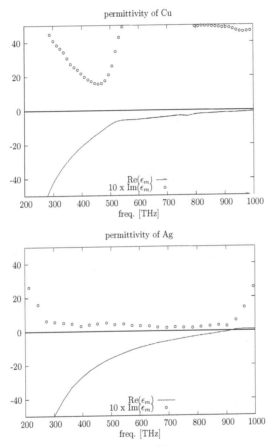

Figure 1. Top: permittivity of Cu; bottom: permittivity of Ag.

3. Modeling techniques

In this section the full wave modeling techniques are introduced and categorized on the basis of their solution method: Finite Elements (FE), Finite Differences in the Time Domain (FDTD), Finite Integration Technique (FIT), and Integral Equations (IE) solved by the Method of Moments (MoM). Based on their theoretical specificities, the application of each method in the case of plasmonics is discussed.

The cradle of computational electromagnetics can be found in the microwave research community. Since this community traditionally is dealing with structures in the order of wavelengths, right from its beginning days, it had no choice than to try to rigorously solve Maxwell's equations. Most standard reference works on full-wave computational electromagnetics by consequence can be found within this community. A history and a comprehensive overview of the different numerical techniques and of their application in computational electromagnetics (CEM) may be found in [2]. Recent work on the last developments in CEM [3] concentrates on the two main approaches of differential and integral methods. A Good review and perspectives concerning the relationship between differential and integral equations (IE) modeling is recommended in a review paper by Miller [4]. A thorough discussion on the different techniques with clarifying examples is also given in [5].

3.1. Differential equation techniques

In time domain, Maxwell's equations are

$$\nabla \times \mathbf{E} = -\frac{d\mathbf{B}}{dt} \tag{1}$$

$$\nabla \times \mathbf{H} = \frac{d\mathbf{D}}{dt} + \mathbf{J} \tag{2}$$

with $\mathbf{E}$ and $\mathbf{B}$ the electric field and the magnetic induction, respectively, $\mathbf{H}$ and $\mathbf{D}$ the magnetic field and the electric induction, respectively, and $\mathbf{J}$ the electric current flowing. In free space, and in a homogeneous, isotropic, time-invariant, linear medium (most materials behave like this in the microwave frequency range), the following relations hold

$$\mathbf{D} = \varepsilon \mathbf{E} \tag{3}$$

$$\mathbf{B} = \mu \mathbf{H} \tag{4}$$

They are called the constitutive relations, with ε and μ the permittivity and permeability of the medium surrounding the observation point considered. It is crucial to point out that in plasmonics in general (3) and (4) cannot be used. For example, the very dispersive properties of metals at optical frequencies (and beyond) prohibit this for these materials.

In time domain, the time variation has to be determined. In frequency domain, it is assumed that all field quantities are varying in a sinusoidal way. This means that the variation with time is known, and only the variation of the fields in space has to be determined. Using complex notation for the frequency domain with a $\exp(j\omega t)$ time dependency, Maxwell's equations become

$$\nabla \times \mathbf{E} = -j\omega \mathbf{B} \tag{5}$$

$$\nabla \times (\mu^{-1}\mathbf{B}) = j\omega \varepsilon \mathbf{E} + \mathbf{J} \tag{6}$$

with $\omega = 2\pi \cdot$ frequency the pulsation. Note that in (5) and (6) ε and μ are used. In general, here they are depending on frequency, and in this way dispersion is fully taken into account (see also section 2.3). In the field of plasmonics, this is an important difference between computational tools in time and in frequency domain. Computational tools in frequency domain may use directly the well-known concepts of permittivity and permeability, albeit that they become frequency dependent. Computational tools in time domain cannot use these concepts directly. As will be shown later, this complicates things.

From a mathematical perspective, Maxwell's equations are differential equations relating vector fields to each other. Differential equation methods in Computational Electromagnetics are methods that directly consider Maxwell's equations (or the Helmholtz wave equations derived from them), with little analytical preprocessing. Basically, these differential equations, which are valid in any point of space and time, are solved by approximating them by difference equations, which are valid in a discrete set of points in space and time. This is done by chopping up space (and time if time domain is considered) in little pieces in which the field variation has a pre-described profile. This reduces the problem of fields varying over space (and time) to a discrete (matrix) problem that can be handled on a computer. The differences between the several differential equation methods are related to the different ways in which space (and time) can be chopped up.

Since the number of unknowns is proportional to the volume and the resolution considered, differential equation methods are particularly suitable for modeling small full three-dimensional volumes that have complex geometrical details, for example smaller closed-region problems involving inhomogeneous media [6]. Intrinsically, differential equations are less suited for open problems. The reason is that in principle they require a discretization of the entire space under consideration. This space is limited in case of closed problems, but corresponds to infinite space in case of open problems. In practice, this problem is solved by the introduction of techniques like Absorbing Boundary Conditions, and Perfectly Matched Layers (PML) [7]. They mimic the wave arriving at the boundary as travelling further to infinity. The quality of these truncating techniques nowadays is very high so that, in practice, the intrinsic problem with open structures has been overcome, albeit it in an approximate numerical way.

The most popular differential equation-based methods are the Finite Element Method (FEM), for example utilized in Ansoft's HFSS software package, and the Finite-Difference Time Domain method (FDTD), which is employed for example by CST's Time Domain transient solver (in the particular case of Cartesian grids), and by Lumerical.

3.1.1. The Finite Element Method (FEM) [7], [8]

FEM is a method based on solving partial differential equations. It is most commonly formulated based on a variational expression. It subdivides space in elements, for example tetrahedra. Fields inside these elements are expressed in terms of a number of basic functions, for example polynomials. These expressions are inserted into the functional of the equations, and the variation of the functional is made zero. This yields a matrix eigenvalue equation whose solution yields the fields at the nodes. FEM gives rise to a very sparse matrix equation, which can be solved using dedicated matrix algebra technology, leading to very fast solution times, considering the huge number of unknowns.

Its first formulations were developed as matrix methods for structural mechanics. This lead to the idea to approximate solids and Courant (1942) introduced an assembly of triangular elements and the minimum of potential energy to torsion problems [9]. The first paper on the application of FEM to electrical problems appeared in 1968 [10]. An extensive review on the history of FEM in electromagnetics was published in an issue of the Antennas and Propagation Magazine [11]. FEM normally is formulated in the frequency domain, i.e. for time-harmonic problems. This means that, as for IE-MoM, the solution has to be calculated for every frequency of interest.

Numerous references can be given developing, explaining, and using FEM. A good book to start with is [8]. A software tool using FEM and very widely spread is Ansoft HFSS.

3.1.2. The Finite-Difference Time-Domain technique (FDTD) [12], [13], [14]

The nature of Maxwell's differential equations is that the time derivative of the H-field is dependent on the curl of the E-field, and the time derivative of the H-field is dependent on the curl of the E-field. These basic properties result in the core FDTD time-stepping relation that, at any point in space, an updated value of an E/H-field in time is dependent on the stored value of the E/H-field and the numerical curl of the local distribution of the H/E-field in space. The numerical translation into a time-stepping algorithm was introduced by Yee in 1966. Indeed, swapping between E-field and H-field updates allows to define a marching-on-in-time process wherein sampled fields of the continuous electromagnetic waves under consideration are used. These waves can be seen to propagate in the Yee lattice, a numerical three-dimensional space lattice comprised of a multiplicity of Yee cells, see Fig. 2. More specifically, Yee proposed a leapfrog scheme for marching-on in time wherein the E-field and H-field updates are staggered so that E-field updates are observed midway during each time-step between successive H-field updates, and vice versa. A huge advantage of FDTD is that this explicit time-stepping scheme avoids the need to solve simultaneous equations, so no matrix inversions are necessary. It also yields dissipation-free numerical wave propagation. Negative is that this scheme results in an upper bound on the time-step to ensure numerical stability. This means that simulations may require many thousands of time-steps for completion. The use of the Yee lattice has proven to be very robust in numerical calculations.

FDTD is extremely versatile since the interaction of an electromagnetic wave with matter can be mapped into the space lattice by assigning appropriate values of permittivity to each electric field component, and permeability to each magnetic field component. This can be done without seriously compromising the speed of the method.

The fact that time is observed directly, and the versatility of the method, make it probably the most efficient technique for complex 3D transient problems. An implementation of the

FDTD technique dedicated to plasmonics and photonics can be found in the commercial tool Lumerical [15].

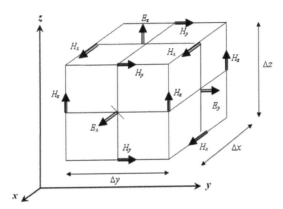

Figure 2. Standard Cartesian Yee cell used in the FDTD technique.

3.1.3. The Finite Integration Technique (FIT) [16]

The Finite Integration Technique was introduced by Weiland in 1977. The word integration does not imply any relation with integral equations. FIT first describes Maxwell's equations on a grid space. The matrix equations for the electromagnetic integral quantities obtained by FIT possess some inherent properties of Maxwell's equations, for example with respect to charge and energy conservation. This makes them very attractive from a theoretical point of view. FIT can be formulated on different kinds of grids, e.g. Cartesian or general non-orthogonal ones, which is a clear advantage. In the time-domain, the resulting discrete grid equations of FIT are, at least in "some" cases, identical to the discrete equations derived with the classical Finite-Difference Time-Domain (FDTD) method. In contrast to FIT, which is applied to the integral form of the field equations, FDTD (as a subset of the finite integration method) is applied to the differential form of the governing Maxwell curl equations. Theoretical links between FDTD and the FIT approach realized in CST may be found in [17]. A comparative study on Time Domain methods was presented in [18]. In some sense, FIT can thus be considered as a powerful generalization of the FDTD technique. A software tool using FIT and very widely spread is CST Microwave Studio.

3.2. Integral Equation techniques (IE)

Integral equation methods make use of Maxwell's equations in integral equation form to formulate the electromagnetic problem in terms of unknown currents flowing on the object to be described. These currents are induced by a field incident on the object. This incident field can be a real incident field, travelling in space, or a bounded wave feeding the object for example via a transmission line. An integral equation solution is fundamentally based on the combination of two equations.

In the case of Electric Field Integral Equations (the case of prime importance to the plasmonics community), the solution is based on considering the electric field. The first equation is

$$\mathbf{E}^{sca}(\mathbf{r}) = \int_{V'} \mathbf{G}(\mathbf{r},\mathbf{r}') \cdot \mathbf{J}^{ind}(\mathbf{r}')dV' \tag{7}$$

which gives the scattered electric field $\mathbf{E}^{sca}$ generated in an arbitrary observation point $\mathbf{r}$ in space in terms of an induced volumetric electric current distribution $\mathbf{J}^{ind}$ flowing within the volume V' of the object considered. The kernel $\mathbf{G}(\mathbf{r},\mathbf{r}')$ is a so-called dyadic Green's function, which is a tensor, and which relates a Dirac impulse type current flowing in $\mathbf{r}'$ to the electric field it generates. The main strength of the integral equation technique is that in many cases the dyadic Green's function can be calculated either directly analytically, such as in homogeneous media, or as some kind of inverse Fourier transform of an analytic function, as is the case for example in layered media. It is crucial to understand that the dyadic Green's function is actually a solution of Maxwell's equations and thus rigorously takes into account the background medium, for example a stack of dielectric layers. This means that unknown currents only have to be assumed on the objects embedded within this background medium. This results in an enormous reduction of unknowns compared to differential equation techniques, which have to model these dielectric layers in just the same way as the objects embedded within them. Basically, the problem formulation automatically covers the entire surrounding space without making any fundamental approximations. As a consequence, the corresponding solution is automatically valid in every point of the background medium. Far field radiation phenomena, surface waves in layered structures, etc., that are vital for efficient and accurate analysis, are analytically included in the solution.

In the case of plasmonics, the second equation is

$$\begin{aligned} \mathbf{J}^{ind}(\mathbf{r}') &= j\omega \left(\varepsilon(\mathbf{r}') - \varepsilon_0\right) \mathbf{E}^{ind}(\mathbf{r}') \\ &= j\omega \left(\varepsilon(\mathbf{r}') - \varepsilon_0\right) \left(\mathbf{E}^{sca}(\mathbf{r}') + \mathbf{E}^{sca}(\mathbf{r}')\right) \end{aligned} \tag{8}$$

It expresses the boundary condition which has to be enforced within the object under consideration. It is a relation between the total field, which is the sum of the incident and the scattered field, and the volumetric electric current, which can be considered as partially being a conduction current (due to the imaginary part of the permittivity), and a polarization current (due to the real part of the permittivity).

Combining (7) and (8) yields the equation from which the currents can be solved. This can be done by chopping up the object considered (only the object, and thus not entire space !!!) in little pieces in which the current variation has a pre-described profile (a so-called basis function). This reduces the problem of currents varying over the object to a discrete (matrix) problem that can be handled on a computer. IE-MoM gives rise to a dense matrix equation, which can be solved using standard matrix algebra technology.

Next to the reduced number of unknowns, there are other theoretical advantages linked to the IE-MoM method. The second advantage is that, if properly formulated, IE-MoM *is variationally stable* since most of the output parameters are expressed in integral form over the equivalent currents. This means that even if the calculated currents differ considerably from the exact solution, integral parameters over both currents may remain very similar. Further, this will be illustrated by showing that even with a rather rough mesh high quality physical results may be obtained. A third advantage is that MoM does not heavily suffer from field singularities for example near sharp edges, since they are analytically incorporated inside the Green's functions. For differential equation methods special care (= a fine mesh) should be taken to describe correctly these field singularities.

The main disadvantage of IE-MoM is that, although there are many efforts in that direction, there is still a lack of matrix solvers operating on dense matrices which are comparable in efficiency to the solvers used in the differential equation techniques (mainly FEM), which yield sparse matrices. This has precluded the use of the very flexible volumetric IE-MoM technique for modeling complex structures on widespread computer systems.

IE-MoM is normally applied in the frequency domain, i.e. for time-harmonic problems. This means that the solution has to be determined at each frequency of interest.

Much more details can be found in the classic and basic Method of Moments (MoM) book [19]. There are many variants of the method. In general, boundary equations can be enforced also at the boundaries of volumes (utilizing Surface Integral Equations (SIE) [20]), next to inside the entire volumes themselves (applying Volume Integral Equations (VIE) [21] at the inside of the components, as described above). Also, the integral equations can be written down in different forms (dyadic form, mixed-potential form, hybrid forms, etc. [22]), which give rise to specific implementations. Further, the first theoretical developments in the field of computational plasmonics are appearing in literature [23].

In the following sections, several aspects linked with the use of integral equation techniques for plasmonic structures are discussed.

3.2.1. 3D volumetric current

The fact that a 3D volumetric current has to be described, as mentioned already, invokes the need for either a real volumetric MoM implementation or the more common surface approach, where the plasmonic component is described by applying the equivalence principle at its surface. The surface approach has already been used, for example in [24]. The volumetric approach as described in [21], has very recently been introduced in the plasmonics community [25]. For realistic structures, where volumes can be embedded within a layer structure, it offers a very flexible solution technique.

3.2.2. Material characteristics

Although strongly varying material characteristics may seem trivial to implement in an integral equation scheme, in some cases more severe consequences occur. For example, some Green's function calculation schemes are based on the fact that the material properties do not change over the frequency band of interest. Any particular implementation of IE-MoM needs this essential relaxation into its formalism, which possibly needs to be adapted.

An advantage of a frequency domain technique like IE-MoM compared to a time domain technique like FDTD is that the ellipsometric measurement data can be used directly in the tool, without any further fitting. FDTD techniques developed for the optical range tend to fit the dielectric response of the metal to the experimentally determined dielectric permittivity using a Drude model [26] or more sophisticated models (for example in Lumerical). However, this may create problems (see further).

3.2.3. Occurrence of layer structures

As explained above, IE technique solvers are formulated making use of Green's functions. These Green's functions can be formulated for multi-layered structures, such that the background medium of the structure may consist of an arbitrary number of horizontal, infinitely stretched, dielectric and metallic layers, which are taken into account analytically. This environment is particularly interesting for plasmonic structures, because they are traditionally deposited on a flat layered dielectric substrate, for example glass. This glass can be contained in the background environment. The only remaining components are local scattering objects with medium or small dimensions compared to the wavelength, such as the dipole in section 5.2.1, but also dots, rods, monomers, dimmers, rings, discs, etc.. With volumetric integral equations these components are replaced by equivalent volume currents, which appear as the primary unknowns in the resulting integral equations. Since the substrates used in normal circumstances are huge compared to the wavelength, edge effects are extremely small and can be neglected. It may thus be concluded that plasmonic structures have specific features which make a IE-MoM solution quite attractive.

4. Software tools

In the following sections, several solvers are briefly described. Several commercial solvers and one academic solver are considered. It has to be emphasized that the author does not claim that this overview is complete. Since the widespread use of computational electromagnetics in plasmonics is quite recent, it is highly probably that there are more solvers that the author is not aware of.

4.1. Commercial software tools

4.1.1. HFSS [27]: FEM

Since it was one of the first tools in the market, and also due to its generality and flexibility, HFFS is one of the tools heavily used in industrial microwave and millimeter wave design environments. The purpose of HFSS is to extract parasitic parameters (S, Y, Z), visualize 3D electromagnetic fields (near- and far-field), and generate SPICE models, all based on a 3D

FEM solution of the electromagnetic topology under consideration. This software is extremely popular and is used for all kinds of purposes. Numerous results for plasmonic topologies can be easily found by googling the words HFSS and plasmonics.

The first step in a new simulation project is to define the geometry of the structure. The geometry is modeled in the GUI or can be imported from another program (AutoCAD, STEP, ...). Then the user continues by defining material properties, boundary conditions, and excitations to the different domains and surfaces of the geometry. If desired one can control the meshing process manually. Finally, the solution parameters are defined. The most important are the solution frequency, the order of the base functions and the matrix solver type. The direct matrix solver is default, but an iterative solver can be used as well. Very useful is the automatic adaptive mesh generation and refinement, which in many cases frees the designer of worrying about which mesh/grid to choose. First a solution for an initial mesh is determined and the solution is assessed. If the solution does not qualify, the mesh is refined and a new solution is computed. This procedure is repeated until one of the exit criteria is fulfilled. It is important to note that HFSS offers no curved elements for a better approximation of curved objects. This means that the mesh along a curved surface has to be chosen very fine. This can be achieved with the 'surface approximation' parameter. Finally the solution, i.e. electric and magnetic fields, currents, and S-parameters can be visualized in 1D, 2D and 3D. All solutions can be exported as files. An 'Optimetrics' toolbox is offered for optimizations, parameter, sensitivity and statistical analysis.

4.1.2. COMSOL multiphysics [28]: FEM

Comsol Multiphysics uses the finite element method to solve partial differential equations in 2D and 3D. It is a multiphysics code, which means that it can handle not only electromagnetics, but also acoustics, mechanics, fluid dynamics, heat transfer, etc.. The geometry, material parameters and boundary conditions have to be set up in the graphical user interface (GUI). Then, after defining boundaries and domains the mesh is generated. This includes the selection of the order of the basis functions and the curved mesh elements (order higher than one means curved element). They are used for a better approximation of curved boundaries. The meshing can be steered by specifying a number of nodes and their distribution on each edge of the model. Comsol Multiphysics offers a wide variety of matrix solvers to solve the system of equations. There are direct solvers, more suitable for smaller problems, and iterative solvers suitable for larger problems. Comsol Multiphysics can generate 2D and 3D plots. Quantities derived from the electromagnetic field, like Poynting vector, energy, and energy loss, are available as predefined variables. Comsol Multiphysics does not allow parameter sweeps or optimization. COMSOL is very easily combined with MATLAB. A simulation set up with the GUI can be saved as a Matlab script, which can be easily modified. The Matlab optimization toolbox can be used to optimize a topology.

4.1.3. JCMsuite [29]: FEM

JCMsuite is a software package based on FEM. It contains dedicated tools for certain simulation problems appearing in nano-optics and plasmonics. It incorporates scattering tools, propagation mode tools, and resonance mode tools. The time-harmonic problems can be formulated in 1D, 2D and 3D. The topologies can be isolated or may occur in periodic patterns, or a mixture of both. Interesting is that dedicated tools for problems posed on cylindrically symmetric geometries are available. The electromagnetic fields are discretized with higher order edge elements. JCMsuite contains an automatic mesh generator (with adaptive features), goal-oriented error estimators for adaptive grid refinement, domain-decomposition techniques and fast solvers.

4.1.4. CST [30]: FIT

CST Microwave Studio (CST MWS) is based on the finite integration technique (FIT). It allows to choose the time domain as well as the frequency domain approach. Despite the presence of transient, eigenmode, and frequency domain solvers within CST MWS, the transient solver is considered as the flag ship module. The Time Domain Solver calculates the broadband behavior of electromagnetic devices in one simulation run with an arbitrarily fine frequency resolution. The modeling of curved structures using the Perfect Boundary Approximation® technique and the modeling of thin perfectly electric conducting sheets with the Thin Sheet Technique® tries to cope with the typical difficulties inherent to FDTD methods for classical structures. Several mesh types can be chosen. The automatic mesh generator detects the important points inside the structure (fixpoints) and locates mesh nodes there. The user can manually add fixpoints on a structure, as well as fully control the number of mesh lines in each coordinate with regards to the specified wavelength. Energy based adaptation of the mesh allows to refine it in a predefined number of passes, providing a mesh refinement of sophisticated design features for the price of longer overall simulation times. Although the FIT in principle can handle material parameters changing over the dielectric volumes defined, this is not implemented yet.

CST, as a general purpose software package being a real competitor for HFSS in the traditional microwave field, has gained a lot of popularity in the last few years. Also for the analysis and design of plasmonic structures, more and more results obtained with CST can be found in literature (just google the words CST and plasmonics). A problem sometimes observed with CST is a ripple in the frequency response in case the tool settings are not appropriate. This is due to the fact that the flagship of CST is inherently a time domain solver.

4.1.5. Lumerical [15]: FDTD

Lumerical is the leading software tool in the plasmonics and photonics community. It is based on the FDTD algorithm, and can be used for 2D and 3D topologies. The topology has to be generated in the GUI or can be imported with GDSII/SEM files. Basic shapes include triangles, rectangular blocks, cylinders, conic surfaces, polygons, rings, user-defined (parametric) surfaces, spheres and pyramids. Several boundary conditions can be used: absorbing (PML), periodic, Bloch, symmetric, asymmetric, and metal boundaries. A non-uniform mesh can be used and automesh algorithms are provided. Since it is a dedicated time domain solver, it has a sophisticated library of Lorentz, Drude, Debye models for the material parameters. Excitation of the structure is possible with waveguide sources, dipoles, plane waves, focused beams and diffraction-limited spots. A scripting language is available to customize simulation. Data can be exported to Matlab or in ASCII.

4.2. Non-commercial software tools

4.2.1. MAGMAS 3D [31]: IE-MoM

MAGMAS 3D is the IE-MoM code developed at the Katholieke Universiteit Leuven, Belgium. It was originally developed in cooperation with the European Space Agency for planar and quasi-planar antenna and scattering structures embedded in a multilayered dielectric background medium and operating in the microwave frequency range. Starting from the topology considered, frequency band needed, and type of excitation, it calculates the network, radiation, and scattering characteristics of the structure under consideration. It is based on a full-wave Mixed-Potential formulation of Electric Field Integral Equations (MP-EFIE), originally applied only to 2D surface currents [22], [32]. New theoretical techniques were developed and implemented within the framework: the Expansion Wave Concept (EWC) [33], [34], the Dipole Modeling Technique (DMT) [35], special de-embedding procedures [36], etc.. In 2007, it was extended with the capability to handle volumetric 3D currents with the VIE technique introduced in [21]. This was crucial in view of its application in plasmonics. In 2009, this led to the first verified simulated results in this field [25]. Now, it is extensively used for plasmonics. The MAGMAS mesh is based on a combination of rectangular and triangular mesh cells. Full mesh control is available in manual meshing mode. A Graphical User Interface is available.

To the best knowledge of the author, to date (June 2012), it is the only IE-MoM framework able to handle arbitrary plasmonic structures embedded in a multilayered environment, with the very general and flexible VIE technique. No similar commercial solvers exist, in contrast to the situation at microwave frequencies, where a multitude of IE solvers can be found.

4.2.2. Other non-commercial tools found in literature

At this moment plasmonics is a mainly experimentally driven research field. Whereas for example in the field of traditional antenna research, numerous papers can be found on modeling as such, this is much less the case in plasmonics, especially in the high impact journals. Most papers are concerned with the description of the physical phenomena occurring and occasionally just mention the numerical tool used.

Most researchers in the plasmonics field use the commercial tools available, and as far as I can see, the main one is Lumerical. In the publications that do treat the numerical modeling of plasmonic structures as such, mainly the FDTD or the FEM technique is applied [37], [38]. Very few papers consider the IE-MoM technique, and if they do, it is the Surface IE technique [39], [40], [41]. In [42] Chremmos uses a magnetic type scalar integral equation to describe surface plasmon scattering by rectangular dielectric channel discontinuities. Even the use of the magnetic current formalism to describe holes, classical at microwave frequencies, already has been used in plasmonics [43]. However, these dedicated developments cannot be categorized under the title „analysis framework" in the sense that they do not allow to handle a wide range of different topologies.

5. Benchmarking

Nowadays, physicists and engineers rely heavily on highly specialized full wave electromagnetic field solvers to analyze, develop, and optimize their designs. Computer-aided analysis and optimization have replaced the design process of iterative experimental modifications of the initial design. It is evident that the underlying solution method for a software tool may significantly influence the efficiency and accuracy by which certain structure types are analyzed. Nevertheless, the commercial focus increasingly switches from such key theoretical considerations to improvements in the area of layout tools and system-level design tools. Therefore, users may get the wrong impression that a given solver is automatically suited to solve any kind of problem with arbitrary precision. This is of course not true.

This section verifies the plausibility of such expectations by presenting a benchmark study for a few plasmonic structures. The study focuses on the capabilities and limitations of the applied EM modeling techniques that usually remain hidden for the user.

5.1. Benchmarking in literature

In literature, not many comparisons between solvers in the field of plasmonics can be found. In [44] Hoffmann et al. consider a single plasmonic topology, a pair of Au spheres, and analyze this structure with several electromagnetic field solvers: COMSOL Multiphysics (FEM), JCMsuite (FEM), HFSS (FEM), and CST Microwave Studio (the FEM solver available in the Studio is used, not the flagship FDTD solver). The output parameter considered is the electric field strength in between the two spheres. Note that all solvers tested are based on the Finite Elements technique in the frequency domain, so it can be expected that the main issues with these solvers for this topology are the same. Very interesting is the fact that in the paper itself CST is categorized as "inaccurate". However, afterwards this was corrected. It seems that wrong material settings were used, and after correction accurate results were obtained. This issue led to the fact that this benchmark example is now presented as a reference example on the CST website.

In [45], several numerical methods are tested for 2D plasmonic nanowire structures: not only the Finite Element Method (FEM) and the Finite Difference Time-Domain (FDTD) technique, but also less „commercial" methods like the Multiple Multipole Program (MMP), the Method of Auxiliary Sources (MAS), and the Mesh-less Boundary Integral Equation (BIE) method are tested. By comparing the results, several conclusions can be drawn about their applicability and accuracy for plasmonic topologies. Differential techniques like FEM and FDTD can reach a high level of accuracy only with a high discretization. In 2D, this is readily affordable on present-day computer systems. There, these techniques have a clear advantage in terms of speed, matrix size, and accuracy. In 3D, due to the rocketing size of the problem, this may not always be that obvious. The advantages may thus disappear in a full 3D analysis of geometrically complicated structures. This happens because matrices become denser or more ill-conditioned. The main conclusion of the paper is that the most efficient method depends on the problem dimension and complexity. This is a similar conclusion as was reached in [46] within the context of the analysis and design of planar antennas.

5.2. Comparison between differential and integral equation techniques

In this section, the differential and integral equation techniques are compared. This is done by choosing a representative solver from each category and using it in the analysis of a basic plasmonic topology. In the category differential techniques, Lumerical is chosen, as it is the most widespread commercial solver in the plasmonic community. Since, as far as we know, there are no commercial integral equation solvers in the plasmonics area, MAGMAS 3D is chosen in this category, the in-house developed solver at Katholieke Universiteit Leuven. The basic topology selected is the plasmonic dipole. It is a structure that can function as a scatterer, but also as a real nano-antenna. The main result of this section is that it draws conclusions on different computational aspects involved in the modeling of nanostructures with the two solvers.

5.2.1. Plasmonic dipole

The plasmonic dipole is depicted in Fig. 3. It is a structure consisting of two nanorods with a gap in between. When the gap is "shortcircuited", it functions as a single rod, when the gap is open, it functions as a device generating an enhanced electric field there. When it is connected to other nanocircuits, it may function as a nano-antenna, both in transmit (Fig. 3b), and in receive (Fig. 3c). The two rods may be fabricated from metals like Au, Ag, Cu, Al, Cr, etc.. In this section the width W and height H of the dipole are kept constant at 40 nm. This value is well-chosen, since it is a value which can be fabricated with sufficient accuracy using present-day nanofabrication technology. The gap width can have different values, depending on the case considered. Note however that 10 nm is about the minimum that can be fabricated with reasonable accuracy nowadays.

5.2.1.1. Quality of the input

It cannot be over-emphasized that Computational Electromagnetics solvers in general produce correct results consistently for a multitude of different structures only if two conditions are met:

1. The user has sufficient general background in the field, and a more specific knowledge about the solution technique implemented in the solver that he is using. If not, there is a huge danger that the solver is not used in a proper way. For example, although this is not necessary to "operate" the solver, if the user has no basic knowledge and does not

realize how the discretization / meshing scheme works, it is impossible to understand and assess the effect of a proper meshing. In the majority of the cases, there will be no problem with that, and automatic meshing schemes will be able to produce good results. However, in unusual and/or challenging cases, often met in scientific research, this issue will be a crucial factor.

2. The user needs to have sufficient knowledge about the actual tool that he is using and its peculiarities, more specific about the way a certain problem has to be imported into the solver. A very good illustration of this has already been addressed in section 5.1 where CST was actually not properly used and produced incorrect results that were published in open literature. Afterwards, this was corrected and the problem disappeared. However, following the proper line of reasoning is not always straightforward. Even experienced researchers sometimes easily can make mistakes. It is the opinion of the author that this is happening too much, even in a lot of peer-reviewed scientific papers.

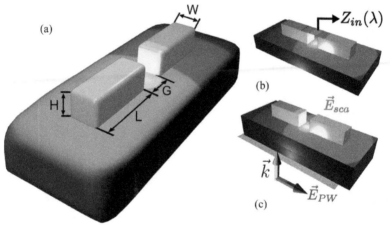

Figure 3. The dipole model studied. a) W and H are set equal to 40 nm and the gap G can vary. b) The dipole as transmitting antenna with a model for the feeding structure located in the gap. c) The dipole as receiving antenna excited by a plane wave.

5.2.1.2. Modeling of the feed

The first parameter studied is the impedance that is seen when the nanodevice operates as a transmitting or receiving antenna. In Fig. 4, the impedance simulated with both MAGMAS and Lumerical is given for an Al dipole of 200 nm length and with a gap of 10 nm. The input impedance is calculated as $Z = V/I$, where Z, V and I are the input impedance, the voltage over the gap G, and the source current, respectively. The exciting source is located in the middle of the gap. MAGMAS uses a physical current filament of finite width w. Lumerical does not have a built-in physically feasible current source model. It uses a (less realistic) pure electrical dipole source, which means that the current I has to be evaluated "manually" by the user based on the magnetic field distribution and Ampere's law. This also means that the impedance will be somewhat depending on the resolution chosen during meshing. It is seen that there is an excellent agreement in the trend of the curves, i.e. the change of input impedance with frequency for different widths of the source current filament in MAGMAS (5 nm and 13.33 nm) is the same as the one in Lumerical. However, in general there is an off-set. Only for a specific width of the current filament in MAGMAS, nl. 5 nm, the two excitation mechanisms become almost completely equivalent and even the offset between the two solvers disappears. A similar agreement was also observed for both gold and silver dipoles.

The main conclusion of this section is that there is still an issue with the modeling of (localized) feeds, especially in Lumerical. The reason is that the plasmonics community is used to consider objects as scatterers, excited by an incident wave. In this case, the models implemented are satisfactory. However, in view of designing nanocircuits, localized feeds and impedances derived from them become important. The basic rule is that the feeding model has to correspond as closely as possible to the actual feed topology that is going to be used later in practice. In this respect, the localized feeding models available in Lumerical today are unsatisfactory.

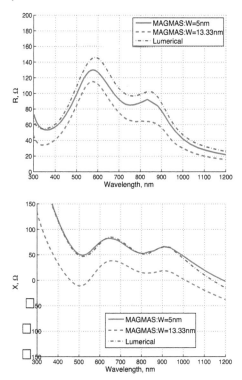

Figure 4. Comparison of impedances obtained with MAGMAS and Lumerical for an aluminum dipole on a glass substrate, L = 200 nm and gap = 10 nm. Two widths w are used in MAGMAS for the source current filament. R and X stand for real and imaginary part.

5.2.1.3. Modeling of losses

The importance of taking into account losses is obvious. Metals at plasmonic frequencies may be very lossy and neglecting this would result in totally erroneous results. However, there is a difference between time domain and frequency domain solvers. Since Lumerical is an FDTD based tool, it works in time domain. This means that it cannot work directly with (complex) permittivities and permeabilities. The material data have to be transformed from functions of frequency into functions of time. This is a very difficult process, involving a convolution, and is subject to various constraints. Local curve fitting of permittivities is very difficult to implement in a time domain solver. Lumerical fits the sampled measurement data with its own multi-coefficient material model, which "provides a superior fit compared to the standard material models (such as Lorentz, Debye, Drude, etc)". However, this produces artefacts, as is illustrated in Fig. 5. There, the relative error is given between the measured imaginary part of the permittivity (which represents losses) and the value obtained for this imaginary part after fitting, for three metals. It is seen that for gold and aluminum, the error is in the order of 20 %, which is already quite elevated. However, in the case of silver, the error reaches a full 100 %, which is unacceptable. In some cases, it will result in a totally wrong prediction for example of the radiation efficiency of nano-antennas [47]. If one forces the material characteristics to be exactly the same in the two solvers, for example by using the interpolated values also in MAGMAS, instead of the real measured ones, the two solvers produce very similar results, see Fig. 6.

The main conclusion is that a highly accurate prediction of losses in Lumerical is still an issue. The material models have to be further refined, especially for silver.

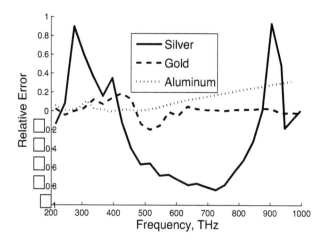

Figure 5. Relative error between interpolated and measured imaginary part of permittivity for silver, gold and aluminum.

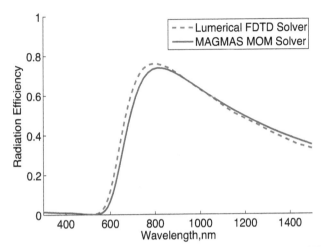

Figure 6. Comparison between FDTD and IE-MoM for a gold dipole in free space, L = 250 nm and the gap is 10 nm. Identical material parameters are used in both solvers.

5.2.1.4. Meshing

It is well known that the mesh quality and resolution are key factors in the accuracy of any solver. The electromagnetic coupling between nearby segments may differ considerably due to the specific meshing used, especially in parts of the structure where rapid topological changes occur. The question is whether the meshes used in the solvers are adequate. This issue was investigated by performing a convergence study in terms of the resolution of the meshes [48].

The extinction cross section of a gold monomer (no gap, or in other words G = 0 nm) calculated using MAGMAS 3D with different meshes is plotted in Fig. 7(a). The analysis of these data demonstrates clearly the stability of IE-MoM. The results obtained even with a very rough mesh 6x1x1 (41.7 *nm* x 40 *nm* x 40 *nm*) provide already a very good estimation of the antenna resonance properties. The differences in extinction cross section calculated with the rough and fine meshes are almost negligibly small. This result is obtained thanks to the variational stability of IE-MoM. On the adjacent figure Fig. 7(b) extinction cross sections are calculated using Lumerical. As expected, in general the results obtained with the FDTD method depend considerably on the chosen mesh. The calculated wavelengths as a function of the mesh cell size are plotted in Fig. 7(c). It should be also noted that in contrast to Fig. 7(a) (IE-MoM) in Fig. 7(b) (FDTD) not only peak positions but also their levels depend clearly on the mesh.

The conclusion is that in Lumerical a fine mesh is mandatory for reliable calculations of the monomer resonant wavelength. Thanks to its variational approach, in IE-MoM, a highly dense mesh is not always needed, especially when scattering problems are studied.

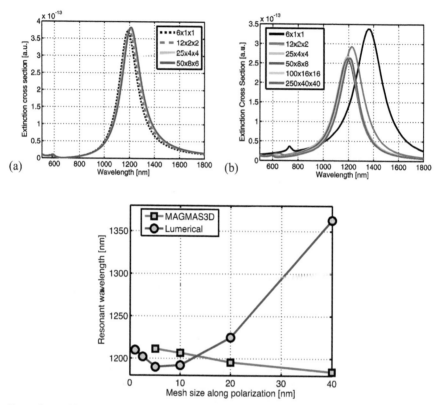

Figure 7. a and b: extinction cross section of a 250 nm x 40 nm x 40 nm gold monomer on a substrate with n=1.5 (a. MAGMAS 3D, b. Lumerical), c. convergence performance of both solvers.

5.2.1.5. Calculation speed

Information on the calculation times for the convergence study of the previous section is given in Table 1. It has to be emphasized that calculation times cannot be directly compared. The computers used were different, and the FDTD tool used 10 processors in parallel. Further, since it is a time domain technique, a calculation time per frequency point cannot be given for FDTD. Nevertheless, combining the data in this table with the data concerning the convergence, given in Fig. 7, it is easily seen that IE-MoM outperforms FDTD for this particular case.

5.2.2. Plasmonic nanojets

In this section it is illustrated what can be reached with computational electromagnetics in the field of plasmonics. The results in this section are extracted from the paper [49], where full details can be found.

IE-MoM		FDTD	
Mesh	Calculation time per frequency point (s)	Mesh	Total calculation time (s) (10 processors in parallel)
6 x 6 x 1	5.4	6 x 1 x 1	38
12 x 2 x 2	10.4	12 x 2 x 2	55
25 x 4 x 4	33.6	25 x 4 x 4	225
50 x 8 x 6	238	50 x 8 x 8	672
		100 x 16 x 16	5893
		250 x 40 x 40	20100

MoM: Intel(R) Xeon(R) CPU E5335 @ 2 GHz, 32 Gb memory
FDTD: d1585g6 4x hc CPU @ 2.8 GHz, 128 Gb memory

Table 1. Calculation times as a function of the mesh for the gold monomer.

Similarly to a pebble hitting a water surface, the energy deposition by femtosecond laser pulses on a gold surface can produce local melting and back-jet. Contrary to water though, a gold surface can be nanopatterned and the excitation of surface plasmon resonances leads to the appearance of hotspots, literally. This was explicitly proven for the first time in a nanopatterned gold surface, composed of a G shaped periodic structure. It was observed there that laser-induced melting and back-jet occur precisely in the plasmonic hotspots. The aid of computational electromagnetics to rigorously analyze the structure by predicting the hotspots was crucial in explaining this phenomenon. Much more details can be found in [49]. It is clear that this type of insight into the basic interaction mechanisms of light with matter at the nanoscale opens up new possibilities for applications.

Another example of what can be reached by applying computational electromagnetics in plasmonics can be found in [50].

Figure 8. Illuminating the sample with femtosecond laser pulses produces a polarization dependent pattern of nanobumps with sharp tips on the sample surface [49]. Left: horizontally polarized light, right: vertically polarized light.

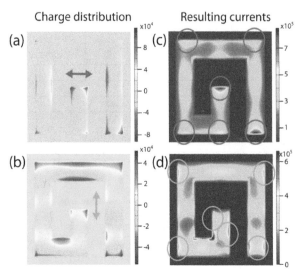

Figure 9. The pattern of melted nanobumps matches that of the plasmonic currents, where Ohmic losses locally increase the temperature [49]. The electric field of light incident on the nanostructures causes the surface charges to oscillate in response to the direction of light polarization: horizontal, in (a), and vertical, in (b). The locations of the resulting electric currents are shown in (c) and (d) respectively. The current maxima are indicated with colored circles matching the pattern of nanobumps in Fig. 8.

6. Conclusions

In this chapter, a brief introductory overview has been given on the use of existing computational techniques and software tools for the analysis and design of plasmonic structures. This field is booming, and many modeling techniques developed at lower frequencies, i.e. in the microwave range, are now being transferred to the plasmonics community, of course with the necessary adaptations. Although the differential equation techniques are the most widespread, both in scientific literature, and in the commercial scene, it is proven that integral equation techniques are a valid alternative. In some cases they are even superior. The author sees a lot of opportunities in this area.

In most cases, the tools are used to analyze structures. Few papers really make dedicated designs. However, this will change in the future. Inevitably, the plasmonics community will follow a similar patch as the traditional antenna community. Whereas 30 years ago, the design of an antenna was based on the accumulated expertise and know-how of many years, nowadays most antennas are designed almost in a single pass by experienced designers using commercial tools. This is possible since the tools in this field have sufficient accuracy and matureness in order to be able to do that. It is my belief that as soon as plasmonics will make the unavoidable shift from the physics to the engineering community, this last community will aim at conceiving applications which will involve designs heavily based on computational electromagnetics.

I would like to end this chapter with the following guideline. The use of two different solvers, based on different theoretical methods (integral and differential) may provide an excellent means to characterize the quality of simulation results. If the two results are in good agreement, it is highly likely that the results are correct. If the two results are in disagreement, a deeper investigation of the structure and its modeling is absolutely necessary.

Author details

Guy A. E. Vandenbosch

Katholieke Universiteit Leuven, Belgium

Acknowledgement

The author gratefully acknowledges the following persons: Zhongkun Ma for the impedance and radiation efficiency comparison between MAGMAS and Lumerical, Dr. Vladimir Volski for the convergence study with MAGMAS and Lumerical, and V. K. Valev for making available the plasmonic nanojet figures.

We also would like to express our gratitude to the KU Leuven Methusalem project and its beneficiary Prof. V. V. Moshchalkov, whose activities created a real stimulus to become personally active in this field, and the fund for scientific research Flanders (FWO-V) for the financial support.

7. References

[1] I. Ahmed, E. H. Khoo, E. Li, and R. Mittra, "A hybrid approach for solving coupled Maxwell and Schrödinger equations arising in the simulation of nano-devices", IEEE Antennas and Wireless Propagation Letters, Vol. 9, pp. 914- 917, 2010.

[2] D. B. Davidson,"A review of important recent developments in full-wave CEM for RF and microwave engineering," IEEE 3rd Int. Conf. Comp. Electromagnetics and Its Applications, pp. PS/1-PS/4, Nov. 2004.

[3] C. W. Townbridge and J. K. Sykulski,"Some Key Developments in Computational Electromagnetics and Their Attribution," IEEE Antennas and Propag. Magazine, vol. 42, no. 6, pp. 503 – 508, Apr. 2006

[4] E. K. Miller,"A Selective Survey of Computational Electromagnetics," IEEE Trans Antennas Propag, vol. 36, no. 9, Sept. 1988, pp. 1281 – 1305

[5] F. Peterson, S. L. Ray, and R. Mittra, "Computational methods for electromagnetics", IEEE Press – Oxford University Press, 1998.

[6] Awadhiya, P. Barba, and L. Kempel, "Finite-element method programming made easy???", IEEE Antennas and Propagation Magazine, vol. 45, pp. 73 – 79, Aug. 2003.

[7] J. M. Jin, The Finite Element Method in Electromagnetics, second edition, John Willey & Sons, Inc., New York, 2002.

[8] J. L. Volakis, A. Chatterjee, and L. C. Kempel, Finite element method for electromagnetics, IEEE Press, Oxford University Press, 1997.

[9] R. L. Courant, "Variational methods for the solution of problems of equilibrium and vibration," Bulletin of the American Mathematical Society, 5, pp. 1-23, 1943.

[10] S. Ahmed, "Finite-element method for waveguide problems," Electronics Letters, vol. 4, Issue 18, pp.387 – 389, Sept. 1968.

[11] R. Coccioli, T. Itoh, G. Pelosi, P. P. Silvester, "Finite-Element Methods in Microwaves: A Selected Bibliography," IEEE Antennas and Propag. Magazine, vol. 38, Issue 6, pp.34 – 48, Dec. 1996.

[12] K. S. Yee, "Numerical Solution of Initial Boundary Value Problems Involving Maxwell's Equations in Isotropic Media," IEEE Trans on Antennas Propag, vol 14, pp. 302-307, 1966.

[13] Taflove, Computational electrodynamics: the finite difference time domain method, Artech House, 1997.

[14] D. M. Sullivan, "Electromagnetic simulation using the FDTD method", Wiley – IEEE Press., 2000, ISBN 978-0-7803-4747-2.

[15] www.lumerical.com

[16] T. Weiland, "A discretization method for the solution of Maxwell's equations for six-component fields", Electronics and Communications AEÜ, vol 31, No. 3, 116–120, 1977.

[17] Bossavit and L. Kettunen,"Yee-like schemes on a tetrahedral mesh, with diagonal lumping," Int. J. Numer. Model., vol. 42, pp. 129 – 142, 1999.

[18] M. Celuch-Marcysiak and W. K. Gwarek, "Comparative study of the time-domain methods for the computer aided analysis of microwave circuits," Int. Conf. on Comp. in Electromagnetics., pp. 30-34, Nov. 1991.

[19] R. F. Harrington, "Field Computation by Moment Methods", New York: Macmillan, 1968.

[20] T. K. Sarkar, E. Arvan, and S. Ponnapalli, "Electromagnetic scattering from dielectric bodies", IEEE Trans. Antennas Propag., vol. 37, pp. 673-676, May 1989.

[21] Y. Schols and G. A. E. Vandenbosch, "Separation of horizontal and vertical dependencies in a surface/volume integral equation approach to model quasi 3-D structures in multilayered media ", IEEE Trans. Antennas Propagat., vol. 55, no. 4, pp. 1086-1094, April 2007.

[22] M. Vrancken and G.A.E. Vandenbosch, Hybrid dyadic-mixed-potential and combined spectral-space domain integral-equation analysis of quasi-3-D structures in stratified media, IEEE Trans Microwave Theory Tech, vol 51, pp. 216-225, 2003.

[23] X. Zheng, V. K. Valev, N. Verellen, Y. Jeyaram, A. V. Silhanek, V. Metlushko, M. Ameloot, G. A. E. Vandenbosch, and V. V. Moshchalkov, "Volumetric Method of Moments and conceptual multi-level building blocks for nanotopologies", Photonics Journal, Vol. 4, No. 1, pp. 267-282 , Jan. 2012.

[24] J. Smajic, C. Hafner, L. Raguin, K. Tavzarashvili, and M. Mishrickey, "Comparison of Numerical Methods for the Analysis of Plasmonic Structures", Journal of Computational and Theoretical Nanoscience, vol. 6, no. 3, pp. 763-774, 2009.

[25] G. A. E. Vandenbosch, V. Volski, N. Verellen, and V. V. Moshchalkov, "On the use of the method of moments in plasmonic applications", Radio Science, 46, RS0E02, doi:10.1029/2010RS004582, 21 May 2011.

[26] R. Qiang, R. L. Chen, and J. Chen, "Modeling electrical properties of gold films at infrared frequency using FDTD method", International Journal of Infrared and Millimeter Waves, vol. 25, no. 8. pp. 1263-1270, 2004.

[27] www.ansoft.com

[28] www.comsol.com

[29] www.jcmwave.com

[30] CST GmbH – Computer Simulation Technology, CST Microwave Studio 2006 user manual, 2006, http://www.cst.com

[31] G. A. E. Vandenbosch, MAGMAS 3D, http://www.esat.kuleuven.be/telemic/antennas/magmas/

[32] G. A. E. Vandenbosch and A. R. Van de Capelle, Mixed-potential integral expression formulation of the electric field in a stratified dielectric medium - application to the case of a probe current source, IEEE Trans. Antennas Propagat., vol. 40, pp. 806-817, July 1992.

[33] F. J. Demuynck, G. A. E. Vandenbosch and A. R. Van de Capelle, The expansion wave concept, part I: efficient calculation of spatial Green's functions in a stratified dielectric medium, IEEE Trans. Antennas Propagat., vol. 46, pp. 397 406, Mar 1998.

[34] G. A. E. Vandenbosch and F. J. Demuynck, The expansion wave concept, part II: a new way to model mutual coupling in microstrip antennas, IEEE Trans. Antennas Propagat., vol. 46, pp. 407-413, March 1998.

[35] B. L. A. Van Thielen and G. A. E. Vandenbosch, Method for the calculation of mutual coupling between discontinuities in Planar Circuits, IEEE Trans. Microwave Theory and Techniques, vol. 50, pp. 155-164, January 2002.

[36] E. A. Soliman, G. A. E. Vandenbosch, E. Beyne and R.P. Mertens, Multimodal Characterization of Planar Microwave Structures, IEEE Trans. Microwave Theory and Techniques, vol. 52, no. 1, pp. 175-182, January 2004.

[37] R. Kappeler, D. Erni, X.-D. Cui, L, Novotny, "Field computations of optical antennas", Journal of Computational and Theoretical Nanoscience, vol. 4, no. 3, pp. 686-691, 2007.

[38] R. Qiang, R. L. Chen, and J. Chen, "Modeling electrical properties of gold films at infrared frequency using FDTD method", International Journal of Infrared and Millimeter Waves, vol. 25, no. 8. pp. 1263-1270, 2004.

[39] B. Gallinet and O. J. Martin, "Scattering on plasmonic nanostructures arrays modeled with a surface integral formulation", Photonics and Nanostructures – Fundamentals and Applications, vol. 8, no. 4, pp. 278-284, 2009.

[40] A. M. Kern and O. J. F. Martin, "Surface integral formulation for 3D simulations of plasmonic and high permittivity nanostructures", J Opt. Soc. Am. A, vol. 26, no. 4, pp. 85-94, 2010.

[41] B. Gallinet, A. M. Kern, and O. J. F. Martin, "Accurate and versatile modeling of electromagnetic scattering on periodic nanostructures with a surface integral approach", J Opt. Soc. Am. A, vol. 27, no. 10, pp. 2261-2271, 2010.

[42] I. Chremmos, "Magnetic field integral equation analysis of surface plasmon scattering by rectangular dielectric channel discontinuities", J Opt. Soc. Am. A Opt. Image Sci. Vis., vol. 27, no. 1, pp. 85-94, 2010.

[43] J. Alegret, P. Johansson, and M. Kall, "Green's tensor calculations of Plasmon resonances of single holes and hole pairs in thin gold films", *New Journal of Physics*, vol. 10, no. 10, 2008.

[44] J. Hoffmann, C. Hafner, P. Leidenberger, J. Hesselbarth, S. Burger, Comparison of electromagnetic field solvers for the 3D analysis of plasmonic nano antennas, Proc. SPIE Vol. 7390, pp. 73900J-73900J-11 (2009).

[45] J. Smajic, C. Hafner, L. Raguin, K. Tavzarashvili, and M. Mishrickey, "Comparison of Numerical Methods for the Analysis of Plasmonic Structures", *Journal of Computational and Theoretical Nanoscience*, vol. 6, no. 3, pp. 763-774, 2009.

[46] Vasylchenko, Y. Schols, W. De Raedt, and G. A. E. Vandenbosch, "Quality assessment of computational techniques and software tools for planar antenna analysis", IEEE Antennas Propagat. Magazine, Feb. 2009.

[47] G. A. E. Vandenbosch and Z. Ma, "Upper bounds for the solar energy harvesting efficiency of nano-antennas", Nano Energy, Vol. 1, No. 3, pp. 494-502, May 2012.

[48] F. Pelayo G. De Arquer, V. Volski, N. Verellen, G. A. E. Vandenbosch, and V. V. Moshchalkov, "Engineering the input impedance of optical nano dipole antennas: materials, geometry and excitation effect", IEEE Trans. Antennas Propagat., Vol. 59, No. 9, pp. 3144-3153, Sep. 2011.

[49] V. K. Valev, D. Denitza, X. Zheng, A. Kuznetsov, B. Chichkov, G. Tsutsumanova, E. Osley, V. Petkov, B. De Clercq, A. V. Silhanek, Y. Jeyaram, V. Volskiy, P. A. Warburton, G. A. E. Vandenbosch, S. Russev, O. A. Aktsipetrov, M. Ameloot, V. V. Moshchalkov, T. Verbiest, "Plasmon-enhanced sub-wavelength laser ablation: plasmonic nanojets", Advanced Materials, Vol. 24, No. 10, pp. OP29-OP35, March 2012.

[50] V. K. Valev, A. V. Silhanek, B. De Clercq, W. Gillijns, Y. Jeyaram, X. Zheng, V. Volskiy, O. A. Aktsipetrov, G. A. E. Vandenbosch, M. Ameloot, V. V. Moshchalkov, T. Verbiest, "U-shaped switches for optical information processing at the nanoscale", Small, Vo. 7, No. 18, pp. 2573-2576, Sep. 2011.

Chapter 2

Modelling at Nanoscale

Paolo Di Sia

Additional information is available at the end of the chapter

1. Introduction

One of the most important aspects at nanoscale concerns the charge transport, which can be influenced by particles dimensions and assumes different characteristics with respect to those of bulk. In particular, if the mean free path of charges, due to scattering phenomena, is larger than the particle dimensions, we have a mesoscopic system, in which the transport depends on dimensions and in principle it is possible to correct the transport bulk theories by considering this phenomenon. A similar situation occurs also in a thin film, in which the smallest nanostructure dimension can be less than the free displacement and therefore requires variations to existing theoretical transport bulk models. Therefore a rigorous knowledge of transport properties is to be acquired. From a theoretical viewpoint, various techniques can be used for the comprehension of transport phenomena, in particular analytical descriptions based on transport equations. Many existing and used theories at today regard numerical approaches, not offering analytical results, which would be of great mathematical interest and in every case suitable to be implemented through the experimental data existing in literature and continuously found by the experimentalists.

To establish the applicability limit of a bulk model and to investigate the time response of systems at nanoscale, recently it has appeared a novel theoretical approach, based on correlation functions obtained by a complete Fourier transform of the frequency-dependent conductivity of the studied system [1]. With this model it is possible to calculate exactly the expressions of the most important connected functions, i.e. the velocities correlation function, the mean free displacement and the diffusion coefficient.

At nanoscale we are in the middle between classical and quantum effects, macroscopic and microscopic properties of the matter [2]. Actually one of the most important experimental technique for the study of the frequency-dependent complex-valued far-infrared photoconductivity $\sigma(\omega)$ is the Time-resolved THz Spectroscopy (TRTS), an ultrafast non-contact optical probe; data are normally fitted via Drude-Lorentz, Drude-Smith or effective medium models [3]. The quantum effects in the nanoworld have opened new ways in a lot of old and new technological sectors.

In the following it is illustrated an overview of the fundamental models that, starting from the Drude model, have attempted to analyze and explain in increasing accuracy the transport phenomena of the matter at solid state level and in particular at nanoscale, until the recent developments concerning the variations of Drude-Lorentz-like models, which involve in particular the concept of plasmon.

2. The Drude model

The most meaningful characteristics of metals are their elevated properties of electric and thermal conduction. Over the years this fact has brought to think in terms of a model in which the electrons are relatively free and can move under the influence of electric fields. Historically two models of the elementary theory of metals are born:

The Drude model, published in 1900 and based on the kinetic theory of an electron gas in a solid. It is assumed that all the electrons have the same average kinetic energy E_m;

The previous model integrated with the foundations of quantum mechanics, called Sommerfeld model.

In the Drude model the valence electrons of atoms are considered free inside the metal; all the electrons move as an electronic gas. It is assumed that all the electrons have the same energy E_m and that it exists a mechanism of collisions among ions and electrons, allowing the thermal equilibrium for the electrons; this fact implies the application of the kinetic theory of gases to such electronic gas. Drude published the theory three years after the discovery of the electron from J. J. Thomson. The free electrons have only kinetic energy, not potential energy; the average energy E_m is therefore $(3/2)k_BT$. It can be correlated to an average quadratic speed v_m from the relation $E_m = (3/2)k_BT = mv_m^2/2$, denoting m the mass of the free electron. At environment temperature it is $v_m \cong 10^7$ cm/s, representing the average thermal speed of the electrons. It was assumed also that the electrons have collisions as instant events, i.e. the time of the diffraction is very smaller with respect to every other considered time. Through such collisions, the electrons acquire a thermal equilibrium corresponding to the temperature T of the metal. If the electrons don't collide, they move in linear way following the Newton laws. The presence of a constant electric field determines an extra average velocity (the drift velocity) given by $v_d = - (eE/m)t$. The relaxation time τ is defined as the average time between two collisions; it is so possible to get a mean free path, defined by $l_{mfp} = v_m \tau$. The current density is $\vec{J} = \sigma_{cond} \vec{E}$, with σ_{cond} electric conductivity. This result, obtained by Drude, is an important goal of the classic theory for the conduction of metals (said "Drude theory").

3. The Drude-Lorentz model

The Lorentz model (1905) is a refining of the Drude model, in which the statistical aspects are specified. The electrons are considered as free charges, with charge "$-e$"; they are described by a maxwellian velocity distribution. Considering an electron gas in a spatial region with a constant electric field, the drift velocity of the electrons is constant; this corresponds to a current density $\vec{j}$ proportional to the applied field $\vec{j}=\sigma_0\vec{E}$, with $\sigma_0=ne^2\tau/m$ (n is the electron density). Estimating the relaxation time τ, Drude and Lorentz have obtained values of conductivity in good accordance with the experiments. In presence of an electric field of the form $E(t)=E_0 e^{-i\omega t}$, the complex conductivity assumes the form $\sigma_\omega=\sigma_0/(1-i\omega\tau)$. Such model, said "Drude-Lorentz model", has received great success, but has also underlined series difficulties.

4. The most utilized Drude-Lorentz-like models

One of the most utilized models for describing experimental transport data is the Drude-Lorentz model [4,5]; with such model the main transport parameters are obtainable. Starting from the Drude-Lorentz model, it is possible to obtain the velocities correlation function, from this the quadratic average distance crossed by the charges as a function of time and examine directly the possible compatibility with the Einstein relation.

Considerable variations of this model were made in this years; the most used are the following:

1. the "Maxwell-Garnett (MG) model": in this model the dielectric function is given by a Drude term with an additional "vibrational" contribution at a finite frequency ω_0 [6], leading to a dielectric function of the form:

$$\varepsilon_{//}(\omega)=1-\frac{\omega_p^2}{\omega(\omega+i/\tau)}+\frac{\omega_s^2}{\omega_0^2-\omega^2-i\gamma\omega} \tag{1}$$

where the amplitude ω_s, the resonant frequency ω_0 and the damping constant γ are material dependent constants. The MG model usually describes an isotropic matrix containing spherical inclusions isolated from each other, such as the metal particles dispersed in a surrounding host matrix.

2. In the "effective medium theories (EMTs)" the electromagnetic interactions between pure materials and host matrices are approximately taken into account [7]. The commonly used EMTs include the Maxwell-Garnett (MG) model and the "Bruggeman (BR) model", which is a variation of the MG model.

Generally, in the THz regime, the dielectric function $\varepsilon_m(\omega)$ consists of contributions of the high-frequency dielectric constant, conduction free electrons, and lattice vibration [8]:

$$\varepsilon_m(\omega)=\varepsilon_\infty-\frac{\omega_p^2}{\omega^2+i\gamma\omega}+\sum_j\frac{\varepsilon_{st_j}\omega_{TO_j}^2}{\omega_{TO_j}^2-\omega^2-i\Gamma_j\omega} \tag{2}$$

in which ε_∞ is the high-frequency dielectric constant, the second term describes the contribution of free electrons or plasmons and the last term stands for the optical phonons.

When the response originates mainly from the contribution of free electrons or plasmons, it is usually adopted the Drude model:

$$\varepsilon_m(\omega)=\varepsilon_\infty -\frac{\omega_p^{\,2}}{\omega^2+i\gamma\omega} \tag{3}$$

which described well the dielectric properties of metals and semiconductors [7].

If instead the interaction of a radiation field with the fundamental lattice vibration plays a dominant role and results in absorption of electromagnetic wave, due to the creation or annihilation of lattice vibration, the dielectric function $\varepsilon_m(\omega)$ mainly consists of the contributions of the lattice vibrations, expressed by the classical pseudo-harmonic phonon model in the first approximation [7]:

$$\varepsilon_m(\omega)=\varepsilon_\infty +\frac{\varepsilon_{st}\,\omega_{TO}^{\,2}}{\omega_{TO}^{\,2}-\omega^2-i\gamma\omega} \tag{4}$$

5. The Smith model

Smith has started from the response theory for the optical conductivity, considering an electric field impulse applied to a system, in order to examine the answer with respect to the current. He has utilized the following Fourier transform for the frequency-dependent complex conductivity:

$$\sigma(\omega)=\int_0^\infty j(t)\exp(i\omega t)dt \tag{5}$$

A field impulse, which exceeds every other acting force on the system, permits to assume the hypothesis that the electrons initially can be considered totally free; therefore it holds:

$$j(0)=n^*e^2\big/m \tag{6}$$

with n^* effective electron density. After calculation, the real part of $\sigma(\omega)$ results:

$$\int_0^\infty \mathrm{Re}\,\sigma(\omega)d\omega=\frac{\pi}{2}j(0)=\frac{\omega_p^2}{8} \tag{7}$$

If the initial current decays exponentially to its initial value with relaxation time τ, it is possible to write:

$$j(t)\big/j(0)=\exp(-t/\tau) \tag{8}$$

from which the standard Drude formula is obtainable:

$$\sigma(\omega)=(n^*e^2\,\tau/m)\big/(1-i\omega\tau) \tag{9}$$

Eq. (9) can be considered the first term of a series of the form:

$$j(t)/j(0)=\exp(-t/\tau)\left[1+\sum_{n=1}^{\infty}c_{n}\left(t/\tau\right)^{n}/n!\right]$$ (10)

The c_{n} factors hold into account of the original electrons speed remained after the n-th collision. The analytical form of the complex conductivity is therefore [9]:

$$\sigma(\omega)=\frac{n^{*}e^{2}\tau}{m(1-i\omega\tau)}\left[1+\sum_{n=1}^{\infty}\frac{c_{n}}{(1-i\omega\tau)^{n}}\right]$$ (11)

6. Plasmonics

6.1. Introduction

A lot of nanostructures, in particular the nanostructured metals, show very complex and interesting optical properties. One of the most interesting and promising phenomena encountered in these structures are electromagnetic resonances, due to collective oscillations of the conduction electrons, said "plasmons". Plasmon modes exist in different geometries and in various metals, in particular in noble metals such as gold and silver. Under determined circumstances, plasmons are excited by light; this leads to strong light scattering and absorption and to an increase of the local electromagnetic field. The interest in plasmon modes has started to the beginning of the 20th century, with Zenneck (1907), Mie (1908), Sommerfeld (1909) and other scientists; recent advances regarding the structure, the manipulation and the observation at nanoscale, have increased the study of this scientific topic. The theoretical efforts have encountered also the growing demand at technological level, in particular for semiconductor based integrated electronic components, optical applications and new nano-components. Actually it remains an important challenge for research and development processes, like the guide of light in integrated optical systems and the interface with electronic components. A lot of nanostructures are believed to be one of the key ingredients of the future optoelectronic devices [10].

6.2. Plasmons

The concept of plasma is very useful in the description of some aspects of the interaction radiation/conductive matter; the free electrons of the conductive material (for example a metal) are considered as an electron fluid with high density, of order of $10^{23}\,cm^{-3}$. Such concept is the base of the classic Drude model, which assumes that the material contains stopped positive ions and a gas of classical not interacting electrons, whose motion results slowed by a force of viscous friction, due to the collisions with the ions, and characterized from a relaxation time τ. The motion of electrons results so casual, due to the continuous collisions with the lattice. The plasma frequency is defined as the proper frequency of the collective motion of electrons in the following way:

$$\omega_p^2 = \frac{\sigma_0}{\varepsilon_0\,\tau} = \frac{ne^2}{\varepsilon_0\,m} \tag{12}$$

with σ_0 conductivity of material and common meaning of the other mathematical symbols. The dielectric constant of metal can be written as a function to the plasma frequency; in first approximation it results:

$$\varepsilon(\omega) = 1 - \frac{\omega_p^2}{\omega^2} \tag{13}$$

From the comparison between ω and ω_p, we can deduce the behaviour of the electromagnetic waves arriving to the metal. If it holds $\omega < \omega_p$, the dielectric constant is negative, therefore its square root is imaginary pure; this involves the reflection of the incident wave. In the contrary case, the square root of the dielectric constant is real and the incident wave can propagate in the medium with a small attenuation.

The plasma oscillations are the fluctuations of charge density, which happen to the frequency ω_p and propagate in the metal. The quanta of such fluctuations inside the volume are the "volume plasmons".

There is also a mechanism on the metal surface, characterized by the "surface plasmons". Normally it appears in the wavelength range between the visible and the infrared for the interface air/metal. Localized surface electromagnetic excitations can exist at the surface of nanoparticles and metallic nanostructures: the "localized surface plasmons" (LSP) [11]. The frequency and the intensity of the radiation are very sensitive with respect to the dimension, form and morphology of the nanostructures [12]. Nanoparticles and nanostructures with smaller dimensions with respect to the wavelength of the exciting light are characterized by a wide absorption band, normally in the range of the visible and near infrared spectrum.

6.3. Related theoretical models

The models concerning plasmonics simulate the extinction of the "localized surface plasmons resonance" (LSPR) from nanoparticles and nanoarrays; the application regards the calculation of the light absorption and the scattering. The most used models are:

- the Mie theory [13];
- the Gans theory [14];
- the discrete-dipole approximation method (DDA) [15];
- the finite-difference time-domain method (FDTD) [16].

6.3.1. The Mie theory

The Mie theory is a theoretical approach concerning the optical properties of the nanoparticles. When the nanoparticle dimension is smaller than the wavelength of the incident light, such theory predicts that the extinction caused by a metallic nanosphere is estimated in the quasi-static and dipole limit. The relative cross section of the process results:

$$\sigma_{ext}=\frac{24\pi N_A a^3 \varepsilon_m^{3/2}}{\lambda \ln(10)}\left[\frac{\varepsilon_i}{(\varepsilon_r+\chi\varepsilon_m)^2+\varepsilon_i^2}\right] \tag{14}$$

where N_A is the surface density, a the radius of the metallic nanosphere, ε_m the dielectric constant of the medium surrounding the nanosphere, λ the wavelength, ε_i and ε_r imaginary and real part of the dielectric function of the metallic nanosphere respectively and χ form factor concerning the nanoparticle. The localized resonance depends also on the interparticle space and on the dielectric constant of the substrate.

6.3.2. The Gans theory

The Gans theory extends the Mie theory to spheroidal particles case. In the case of polarization of incident light parallel to the symmetry axis of the spheroid, the cross section of the process is given by:

$$\sigma_{ext}=\frac{2\pi V \varepsilon_m^{3/2}}{3\lambda}\sum_j\left[\frac{(1/P_j)^2\varepsilon_j}{(\varepsilon_r+\frac{1-P_j}{P_j}\varepsilon_m)^2+\varepsilon_j^2}\right] \tag{15}$$

with V volume of nanoparticle and P_j depolarization factors along the three cartesian aces, which hold into account of the anisotropic form of particles.

6.3.3. The discrete-dipole approximation method (DDA)

With this method the nanoparticles are divided in a cubic array of N polarizable dipoles, with polarizabilities α_i determined by the dielectric function of nanoparticles. The induced dipole P_i of every element results $P_i=\alpha_i E_{loc,i}$ in an applied plane wave field. The cross section is determined by:

$$\sigma_{ext}=\frac{4\pi k}{\left|\vec{E}_{inc}\right|^2}\sum_{j=1}^{N}\mathrm{Im}\left(\vec{E}_{inc,j}^{*}\cdot\vec{P}_j\right) \tag{16}$$

with $k=\omega/c$ (in vacuum). Such method can be applied for the calculation of the absorption, scattering, extinction and other optical properties of nanoparticles of various forms and dimensions. It is considered one of the most important methods for the understanding of the structural characteristics and optical properties of nanomaterials and nanostructures.

6.3.4. The finite-difference time-domain method (FDTD)

This is a method of numerical calculation in order to resolve the Maxwell equations directly in the time domain. Being a time-domain method, the solutions can cover a wide frequency range with a single process of simulation. It is a versatile and useful technique in applications where the resonance frequencies are not exactly known. A great variety of magnetic and dielectric materials can be modelled in relatively simple way with such method [17].

7. Linear response theory: a new interesting idea

We start considering a system with an hamiltonian of the form:

$$H = H_0 + H_1 \tag{17}$$

with H_1 having small effects with respect to H_0, and of the form:

$$H_1 = \lambda A e^{-i\omega t} e^{\eta t} \tag{18}$$

being λ a real quantity and η positive, so that in remote past the perturbation is negligible (adiabatic representation: $\lim_{t \to -\infty} H_1 = 0$). In the case of an electric field of frequency ω we have:

$$H_1 = e \vec{E} \cdot \vec{r} \tag{19}$$

If the electric field is constant in space and its evolution depends on time, it is writable as follows:

$$\vec{E} = \vec{E}_0 e^{i\omega t} \tag{20}$$

The time dependent corresponding current is:

$$\vec{J}(t) = \sigma(\omega) \vec{E}(t) \tag{21}$$

The conductivity $\sigma(\omega)$ is in general a complex function of the frequency ω and can be deduced from linear response theory. Following the standard time-dependent approach, it is possible to find a general formula for the linear response of a dipole moment density $\vec{B} = e\vec{r}/V$ in the β direction with the electric field $\vec{E}$ in the α direction, where V is the volume of the system. This permits to deduce the susceptivity $\chi(\omega)$, which is correlated to $\sigma(\omega)$ through the relation:

$$1 + 4\pi\chi(\omega) = 1 + 4\pi i \frac{\sigma(\omega)}{\omega} \tag{22}$$

Analytical calculations permit to write the real part of the complex conductivity $\sigma(\omega)$ as:

$$\operatorname{Re}\sigma_{\beta\alpha}(\omega)=\frac{e^2}{V}\frac{\omega\pi}{\hbar}S_{\beta\alpha}(\omega)\left(1-e^{-\hbar\omega/KT}\right) \tag{23}$$

with:

$$S_{\beta\alpha}(\omega)=\int_{-\infty}^{+\infty}dt\left\langle \vec{r}^{\alpha}(0)\,\vec{r}^{\beta}(t)\right\rangle_T e^{-i\omega t} \tag{24}$$

The part $\left\langle .....\right\rangle_T$ in the integral (24) is the thermal average, and the exponential factor arises from equilibrium thermal weights for Fermi particles. By considering the identity

$$v=\frac{d}{dt}r=\frac{i}{\hbar}\left[H,r\right],$$

Eq. (23) can be written in a form containing the velocities correlation function instead of the position correlation function. Assuming the high temperature limit $\hbar\omega\ll KT$ (valid in such contests), we obtain:

$$\operatorname{Re}\sigma_{\beta\alpha}(\omega)=\frac{e^2}{2VKT}\int_{-\infty}^{+\infty}dt\left\langle \vec{v}^{\alpha}(0)\vec{v}^{\beta}(t)\right\rangle_T e^{-i\omega t} \tag{25}$$

The integral in Eq. (25) spans the entire t axis, so we can perform the complete inverse Fourier transform of this equation. It gives:

$$<\vec{v}^{\alpha}(0)\,\vec{v}^{\beta}(t)>_T=\frac{KTV}{\pi e^2}\int_{-\infty}^{+\infty}d\omega\,\operatorname{Re}\sigma_{\beta\alpha}(\omega)e^{i\omega t} \tag{26}$$

The new introduced key idea is the possibility to perform a complete inversion of Eq. (26) on temporal scale, i.e. considering the entire time axis ($-\infty$, $+\infty$), not the half time axis (0, $+\infty$), as usually considered in literature [9,18]. This idea is viable if we consider the real part of the complex conductivity $\sigma(\omega)$. Via contour integration by the residue theorem in Eq. (26), the integral is determined by the poles of $\operatorname{Re}\sigma(\omega)$. This leads to an exact formulation and gives a powerful method to describe the velocities correlation function (and consequently the mean square displacement and the diffusion coefficient) in analytical way.

8. The other important functions in the nano-bio-context

Another interesting quantity at nanolevel is the mean squared displacement of particles at equilibrium, defined as:

$$R^2(t)=\left\langle [\vec{R}(t)-\vec{R}(0)]^2\right\rangle \tag{27}$$

where $R(t)$ is the position vector at time t. Through a coordinate transformation relative to the integration region it is possible to rewrite the relation (27) as follows:

$$R^2(t)=2\int_0^t dt'\left(t-t'\right)\left\langle\vec{v}(t')\cdot\vec{v}(0)\right\rangle \tag{28}$$

The mean squared displacement can therefore be evaluated through the velocities correlation function. Through Eq. (26) it is possible to deduce also the diffusion coefficient D:

$$D(t)=\frac{dR^2(t)}{2dt}=\int_0^t dt'\left\langle\vec{v}(t')\cdot\vec{v}(0)\right\rangle \tag{29}$$

The three relations (26), (28), (29) are fundamental in deducing the most important characteristics concerning the transport phenomena.

9. About the complex conductivity $\sigma(\omega)$

Let us consider a particle in a region in which there is an electric time-oscillating field, direct along z-axis, with an elastic-type and a friction-type force acting on system; the dynamic equation of the particle can be written as:

$$m\ddot{z}=-kz-\lambda\dot{z}+eE_oe^{-i\omega t} \tag{30}$$

where m is the mass of the particle, k the strenght constant of the oscillators, $\lambda=m/\tau$ and τ is the relaxation time. We consider solutions of the form:

$$z(t)=z_0e^{-i\omega t} \tag{31}$$

The current density in the field direction is:

$$j_z=ne\dot{z}=nez_0(-i\omega)e^{i\omega t}=\sigma E \tag{32}$$

From Eq. (32), via analytical calculations, it is possible to obtain the frequency dependent complex conductivity:

$$\sigma=\frac{i\omega ne^2}{m(\omega^2-\omega_0^2+i\lambda\omega)} \tag{33}$$

The real part of the complex conducivity (33) results [1,19]:

$$\mathrm{Re}\,\sigma(\omega)=\frac{\tau ne^2}{m}\frac{\omega^2/\tau^2}{\left(\omega_0^2-\omega^2\right)^2+\omega^2/\tau^2} \tag{34}$$

10. About the Drude-Lorentz-like models: a new promising "plasmon model"

Recently it has been published a new formulation of the Drude-Lorentz model [1,20], based on linear response theory and resonant plasmonic mode; it is able to accommodate some of the observed departures and gives a detailed description of the dynamic response of the carriers at nano-level. One of the peculiarities of this new model is the inversion of relation (26), extending the integration on time to the entire time axis (-∞, +∞); this procedure is not trivial and it was introduced for the first time in the nano-bio-context. One of the main advantages is the possibility to obtain analytical relations for the description of the dynamic behaviour of nanosystems. Such relations are functions of parameters, experimentally obtainable through Time-resolved techniques. An important consequence of the application of such formulation turns out to be the mathematical justification of the unexpected experimental results of high initial mobility of charge carriers in devices based on semiconductor nanomaterials, as the dye sensitized solar cells (DSSCs). The increase of the diffusion coefficient implies a raise of the efficiency of such electro-chemical cells. The new formulation is able to describe very adequately the properties of transport of nano-bio-materials, but it holds also for other objects, like ions [19].

Many experimental data have indicated that plasmon models describe nanostructured systems in particularly effective way. At quantum level, the key factors incorporating the quantum behaviour are the weights f_i, related to plasma frequencies by:

$$\omega_{p_i}^2 = \frac{4\pi N e^2}{m} f_i \tag{35}$$

where N is the carrier density, m and e respectively the mass and the charge of the electron.

The velocities correlation function in quantum case assumes the form:

$$\langle \vec{v}(0) \cdot \vec{v}(t) \rangle = \left(\frac{KT}{m} \right) \sum_i \left(f_i \exp\left(-\frac{t}{2\tau_i} \right) \left[\cos\left(\frac{\alpha_{iR}}{2} \frac{t}{\tau_i} \right) - \frac{1}{\alpha_{iR}} \sin\left(\frac{\alpha_{iR}}{2} \frac{t}{\tau_i} \right) \right] \right) \tag{36}$$

with:

$$\alpha_{iR} = \sqrt{4\tau_i^2 \omega_i^2 - 1} \; ; \tag{37}$$

$$\langle \vec{v}(0) \cdot \vec{v}(t) \rangle = \frac{1}{2} \left(\frac{KT}{m} \right) \sum_i \left(\left(\frac{f_i}{\alpha_{iI}} \right) \left[(1+\alpha_{iI}) \exp\left(-\frac{(1+\alpha_{iI})}{2} \frac{t}{\tau_i} \right) - (1-\alpha_{iI}) \exp\left(-\frac{(1-\alpha_{iI})}{2} \frac{t}{\tau_i} \right) \right] \right) \tag{38}$$

with:

$$\alpha_{iI} = \sqrt{1 - 4\tau_i^2 \omega_i^2} \; . \tag{39}$$

(K is the Boltzmann's constant, T the temperature of the system, ω_i and τ_i frequencies and decaying times of each mode respectively).

Integrating these expressions through Eqs. (28) and (29), we obtain the mean squared displacement R^2 and the diffusion coefficient D [20,21].

11. Some interesting results of the new model

The model has demonstrated at today high generality and good accordance with experiments. It has been applied in relation to the most commonly used and studied materials at nano-level, i.e. zinc oxide (ZnO), titanium dioxide (TiO₂), gallium arsenide (GaAs), silicon (Si) and carbon nanotubes (CN), also at nano-bio-sensoristic level [19,22].

In Figure 1 it is showed R^2 for doped Silicon. For this systems the conductivity is the contribution of a Drude-Lorentz term and a Drude term. At large times the Drude-Lorentz term leads to an R^2 approaching a constant value (Figure 2), while the Drude term alone is the dominant one. Therefore, for sufficiently large times, only the Drude term survives.

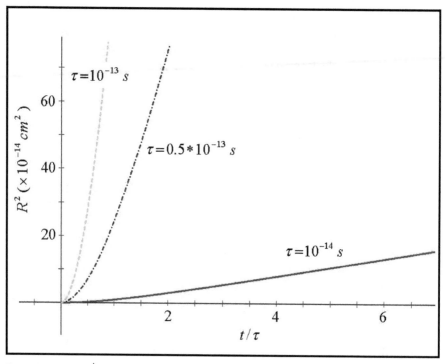

Figure 1. R^2 vs $x = t/\tau$ for some representative values of τ, typical of doped Silicon [23] ($\omega_0 = 0$, T=300K) (Drude model). A complete description of R^2 for Silicon requires the evaluation of the contribution of the Drude-Lorentz part.

We can observe that the linear relation at large times becomes quadratic at smaller times. The cross-over between the two regimes occurs at times comparable to the scattering time.

This means that diffusion occurs after sufficient time has elapsed, so that scattering events become significant, while at smaller times the motion is essentially ballistic.

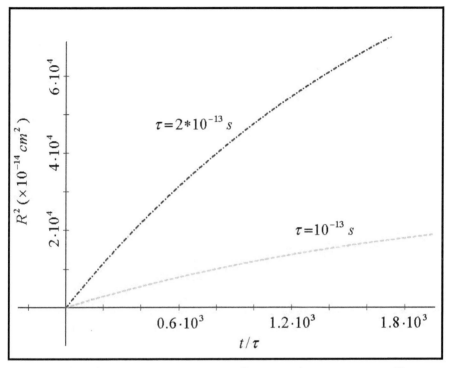

Figure 2. R^2 vs $x = t/\tau$ for 2 values of τ ($\omega_0 = 1.12 \cdot 10^{11}$ Hz dot-dashed line; $\omega_0 = 2.24 \cdot 10^{11}$ Hz dashed line) for TiO$_2$ ($m=6m_e$, $T=300$K). Saturation values occur at sufficiently large t.

The Drude behaviour is contrasted by the plasmon behaviour. It has reported representative cases at large times $x = t/\tau \gg 1$ (Figure 2) for TiO$_2$ nanoparticle films with small frequencies ω_0, i.e. close to the Drude case. It is observable that all of the curves reach a plateau value at sufficiently long times and that the slope within a given time interval increases with τ. As consequences of this behaviour, the plateau of R^2 may become larger than the size of the nanoparticles composing the films, depending on the parameters ω_0 and τ, indicating that carriers created at time $t = 0$ will have enhanced mobility in the nanoporous films at small times, on the order of few τ, in contrast with a commonly expected low mobility in the disordered TiO$_2$ network. Secondly, the time derivative of R^2 decreases to zero at very large times, indicating localization by scattering and therefore leading to decreasing mobility and to the existence of a strictly insulating state for t→∞

Few cases corresponding to short time behaviour and large frequencies are showed in Figure 3, with the scattering time in the range 10^{-13}-10^{-14}s.

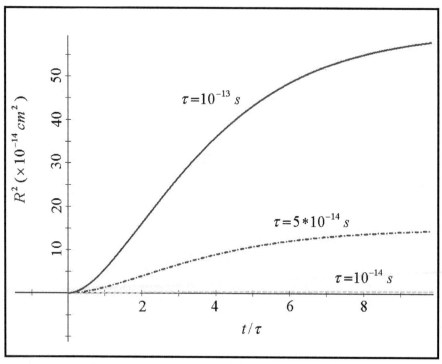

Figure 3. R^2 vs $x = t/\tau$ for 3 values of τ ($\omega_0 = 0.5 \cdot 10^{13}\,Hz$ solid line; $\omega_0 = 10^{13}\,Hz$ dot-dashed line; $\omega_0 = 0.5 \cdot 10^{14}\,Hz$ dashed line) for TiO$_2$ (m=6m_e, T=300K).

In relation to the behaviour of the velocities correlation function, excluding the Drude model ($\omega_0 = 0$), characterized by simple exponential decreasing functions with time, the correlation function of velocities corresponds to either a damped oscillatory behaviour (Eq. (36)), or to a superposition of two exponentials (Eq. (38)), with quite different decay times depending on the value of α_I (Figures 4, 5).

At $\alpha \cong 1$, i.e. close to Drude behaviour, proper case of some systems [3], the current results the superposition of a short $\approx \tau$ and long $\tau/(1-\alpha_I)$ time decay modes. The current is a damped oscillating function of time (Eq. (36)) when $\omega_0 \tau \gg 1$, and will have double exponential behaviour for $\omega_0 \tau \ll 1$ (Eq. (38)).

From this analysis, it is viable the possibility that the previous results can give an explanation of the ultra-short times and high mobilities, with which the charges spread in mesoporous systems, of large interest in photocatalitic and photovoltaic systems [24]; the relative short times (few τ), with which charges can reach much larger distances than typical dimensions of nanoparticles, indicate easy charges diffusion inside the nanoparticles. The unexplained fact, experimentally found, of ultrashort injection of charge carriers (particularly in Grätzel's cells) can be related to this phenomenon.

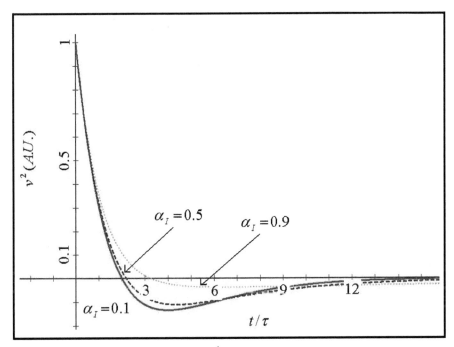

Figure 4. Velocities correlation function vs $x = t/\tau$ for some values of α_I ($m=6m_e$, T=300K).

In systems with charge localization, where the carrier mean free path is comparable to the characteristic dimension of the nanoparticles, the conductivity response becomes more complicated. The response at low frequencies is characterized by an increasing real part of the conductivity and by a negative imaginary part. Deviations from the Drude model become strong in nanostructured materials, such as photoexcited TiO_2 nanoparticles, ZnO films, InP nanoparticles [25], semiconducting polymer molecules [26], and carbon nanotubes [27].

In isolated GaAs nanowires the electronic response exhibits a pronounced surface plasmon mode, that forms within 300 fs, before decaying within 10 ps as a result of charge trapping at the nanowire surface. The conductivity in this case was fitted by using the Drude model for a plasmon and the mobility found to be remarkably high, being roughly one-third of that typical for bulk GaAs at room temperature.

The Smith model with $c_1 = -1$ is obtained as a limit of the new introduced plasmon model when $\alpha_I \to 0$. On performing this limit in Eq. (38) one finds the expression obtained by Smith with a scattering time twice the plasmon scattering time τ. The situation $\alpha_I \to 0$ corresponds to $\tau \approx 1/2\omega_o$. From the other hand, both Smith's and the new model reduce to the Drude model in the limit $\omega_o \to 0$. So, although the two models are analytically different, their predictions are expected to be quite similar. From Figure 4 it is possible to note that, for

$\alpha_I \to 0$, the lower curve is of the form of the Smith curve. The backscattering mechanism invoked by Smith arises in a natural way in this new model, without further assumptions on successive scattering events.

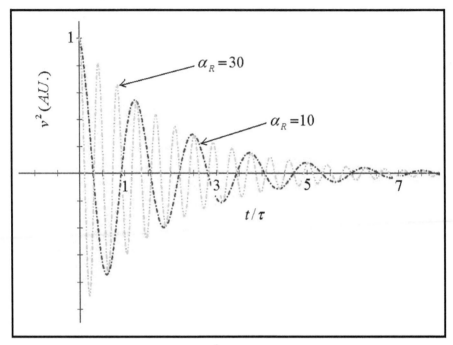

Figure 5. Velocities correlation function vs $x = t/\tau$ for two values of α_R (m=6m_e, T=300K). Evident exponentially damped oscillations are displayed in this case.

Time evolution of charges can be systematically studied by means of time resolved techniques. The THz technique allows a detailed investigation of very short time behaviour, such as the photoinjection, as well as longer time behaviour such as thermalized motion. The studies have indicated a common mechanism of the short time domain in nanostructures, where carriers are close to Drude behaviour with a rather large diffusion coefficient, followed by a range with time decreasing mobility. This latter stage is characterized by decay of the response as the superposition of a short time and a longer time exponentials.

The time response of ZnO films, nanowires, and nanoparticles to near-UV photoexcitation has been investigated in THz experiments. Films and nanoparticles show ultrafast injection, but with the addition of a second slower component. For ZnO nanoparticles, double exponential decay indicates characteristic times $\tau_1 \approx 94\,ps$ and $\tau_2 \approx 2.4\,ns$. From Eq. (38) it is possible to deduce a ratio of two different relaxation times τ_1 and τ_2; it is $\tau_1 / \tau_2 = (1-\alpha_I)/(1+\alpha_I)$. Therefore, on using the experimental values, we find $\alpha_I = 0.92$, from which we deduce $\omega_0 \tau \cong 0.2$. For τ of the order of 10^{-13}s, we obtain $\omega_0 \cong 1.5 \cdot 10^{12}\,Hz$, which is of the correct order of magnitude for the resonance in the infrared. The same procedure for injection times of films (500 nm grains) leads to $\omega_0 \cong 2.5 \cdot 10^{12}\,Hz$ (for $\tau \cong 10^{-13}\,s$).

In GaAs nanowires [27,28] a resonance at $\omega_0 = 0.3 - 0.5\,THz$ has been suggested to explain the frequency-dependent conductivity obtained from THz experiments. A long characteristic obtained time $\tau_c \approx 1.1 \div 5 \cdot 10^{-12}\,s$ is in accordance with the new model, corresponding to $\tau_c = \tau / (1 - \alpha_I) = 5.10^{-12}\,s$ and leading to $\alpha_I = 0.1 \div 0.8$ for $\tau = 10^{-12}\,s$ and $\omega_0 \tau = 0.3 \div 0.49$.

Similar considerations can be applied to single walled carbon nanotubes (SWCN), where the resonance state is at $\omega_0 = 0.5\,THz$ with Drude-Lorentz scattering time $\tau \approx 10^{-13}$ s and to sensitized TiO_2 nanostructured films. In the latter the conductivity of electrons injected from the excited state of the dye molecule into the conduction band of TiO_2 is initially more Drude-like, and then evolves into a conduction dominated by strong backscattering as the carriers equilibrate with the lattice. These results conform to the backscattering mechanism of the new plasmon model.

In Figure 6 it is presented the fitting of data of GaAs photoconductivity [27,28] with the new model.

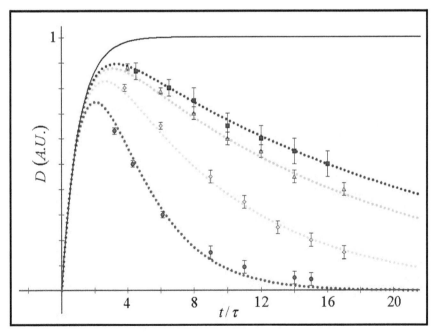

Figure 6. Behaviour of the diffusion coefficient D for α_I (Eq. (39) at classical level [1]) varying in the interval [0.1-1] and $\tau = 10^{-12}$ s. Note that $D \rightarrow 0$ for $t \rightarrow \infty$, indicating absence of diffusion at very long times.

In this figure the upper curve is the Drude result corresponding to $\omega_0 = 0$. It is the highest value of the diffusion coefficient, as ω_0 varies from zero to its maximum value. The dots in Figure 6 are experimental data derived from THz spectroscopy, the error bars on dots represent the distribution of the experimental data and the lines the predictions of the new plasmon model. D tends to a vanishing value as $t \to \infty$, the larger ω_0 the faster this vanishing occurs. The results can be interpreted as follows: at early time, of the order of τ, the system behaves as Drude-like, irrespective of ω_0, with carriers assuming large mobility values (the numerical evaluation gives $D \approx 7.5$ cm^2/s for the case of Figure 6); at increased times, charges become progressively localized as a result of scattering, significantly in agreement with the conclusions drawn by THz spectroscopy [29,30].

12. Conclusions

It is hoped that this new plasmon model will be useful to describe experimental data as an alternative to other generalizations of the Drude model. The principal findings of it are the reversal of the current in small systems like nanoparticles and the time dependence of the transport parameter D, which describes the dynamics of the system at short and long time, with increased mobilities at very short times followed by localization at longer times. This unusual behaviour, with respect to the predicted Drude like, converges as a whole in the peculiar frequency dependence of the optical conductivity. The strength of the new model consist of its ability to accommodate this behaviour and include previous models, like the Smith model [19].

Author details

Paolo Di Sia

Free University of Bozen-Bolzano, Bruneck-Brunico (BK), Italy

13. References

[1] Di Sia P. An Analytical Transport Model for Nanomaterials. Journal of Computational and Theoretical Nanoscience 2011; 8: 84-89.

[2] Wolf EL. Nanophysics and Nanotechnology: An Introduction to Modern Concepts in Nanoscience. John Wiley & Sons; 2006.

[3] Schmuttenmaer CA. Using Terahertz Spectroscopy to Study Nanomaterials. Terahertz Science and Technology 2008; 1(1): 1-8.

[4] Ziman M. Principles of the Theory of Solids. Cambridge University Press; 1979.

[5] Kittel C. Introduction to Solid State Physics. Wiley New York; 1995.

[6] Levy O, Stroud D. Maxwell Garnett theory for mixtures of anisotropic inclusions: Application to conducting polymers. Physical Review B 1997; 56(13): 8035-8046.

[7] Han J, Zhang W, Chen W, Ray S, Zhang J, He M, Azad AK, Zhu Z. Terahertz Dielectric Properties and Low-Frequency Phonon Resonances of ZnO Nanostructures. The Journal of Physical Chemistry C 2007; 111(35): 13000-13006.

[8] Han JG, Wan F, Zhu ZY, Liao Y, Ji T, Ge M, Zhang ZY. Shift in low-frequency vibrational spectra of transition-metal zirconium compounds. Applied Physics Letters 2005; 87(17): 172107-172109.

[9] Smith NV. Classical generalization of the Drude formula for the optical conductivity. Physical Review B 2001; 64(15): 155106-155111.

[10] Shipway AN, Katz E, Willner I. Nanoparticle arrays on surfaces for electronic, optical and sensoric applications. ChemPhysChem 2000; 1: 18-52.

[11] Hutter E, Fendler JH. Exploitation of Localized Surface Resonance. Advanced Materials 2004; 16(19): 1685-1706.

[12] Willets KA, Van Duyne RP. Localized Surface Plasmon Resonance Spectroscopy and Sensing. Annual Review of Physical Chemistry 2007; 58: 267-297.

[13] Stuart DA, Haes AJ, Yonzon CR, Hicks EM, Van Duyne RP. Biological applications of localised surface plasmonic Phenomenae. IEE Proceedings-Nanobiotechnology 2005; 152(1): 13-32.

[14] Gulati A, Liao H, Hafner JH. Monitoring Gold Nanorod Synthesis by Localized Surface Plasmon Resonance. The Journal of Physical Chemistry B 2006; 110(45): 22323–22327.

[15] Zhang ZY, Zhao Y-P. Optical properties of helical Ag nanostructures calculated by discrete dipole approximation method. Applied Physics Letters 2007; 90: 221501-221503.

[16] Sherry LJ, Chang S-H, Schatz GC, Van Duyne RP. Localized Surface Plasmon Resonance Spectroscopy of Single Silver Nanocubes. Nano Letters. 2005; 5(10): 2034–2038.

[17] Zhang W, Yue Z, Wang C, Yang S, Niu W, Liu G. Theoretical models, Fabrications and Applications of Localized Surface Plasmon Resonance sensors. 978-1-4244-4964-4/10/$25.00©2010 IEEE 2010; pp. 4; Project supported by the National Natural Science Foundation of China (Grant N. 60574091, 60871028).

[18] Sassella A, Borghesi A, Pivac B, Pavesi L. Characterization of porous silicon by microscopic Fourier transform infrared spectroscopy; 266-267. 9th International Conference on Fourier transform spectroscopy; 1994.

[19] Di Sia P. Classical and quantum transport processes in nano-bio-structures: a new theoretical model and applications. PhD Thesis. Verona University – Italy; 2011.

[20] Di Sia P. An Analytical Transport Model for Nanomaterials: The Quantum Version. Journal of Computational and Theoretical Nanoscience 2012; 9: 31-34.

[21] Di Sia P. Oscillating velocity and enhanced diffusivity of nanosystems from a new quantum transport model. Journal of Nano Research 2011; 16: 49-54.

[22] Di Sia P. New theoretical results for high diffusive nanosensors based on ZnO oxides. Sensors and Transducers Journal 2010; 122(1): 1-8.

[23] Pirozhenko I, Lambrecht A. Influence of slab thickness on the Casimir force. Physical Review A 2008; 77: 013811-013818.

[24] Grätzel M. Solar Energy Conversion by Dye-Sensitized Photovoltaic Cells. Inorganic Chemistry 2005; 44(20): 6841-6851.

[25] Nienhuys H-K, Sundström V. Influence of plasmons on terahertz conductivity measurements. Applied Physics Letters 2005; 87: 012101-012103.

[26] Hendry E, Koeberg M, Schins JM, Nienhuys HK, Sundström V, Siebbeles LDA, Bonn M. Interchain effects in the ultrafast photophysics of a semiconducting polymer: THz time-domain spectroscopy of thin films and isolated chains in solution. Physical Review B 2005; 71: 125201-125210.

[27] Parkinson P, Lloyd-Hughes J, Gao Q, Tan HH, Jagadish C, Johnston MB, Herz LM. Transient Terahertz Conductivity of GaAs Nanowires. Nano Letters 2007; 7(7): 2162–2165.

[28] Parkinson P, Joyce HJ, Gao Q, Tan HH, Zhang X, Zou J, Jagadish C, Herz LM, Johnston MB. Carrier Lifetime and Mobility Enhancement in Nearly Defect-Free Core–Shell Nanowires Measured Using Time-Resolved Terahertz Spectroscopy. Nano Letters 2009; 9(9): 3349–3353.

[29] Di Sia P, Dallacasa V, Dallacasa F. Transient conductivity in nanostructured films. Journal of Nanoscience and Nanotechnology 2011; 11: 1-6.

[30] Di Sia P, Dallacasa V. Anomalous charge transport: a new "time domain" generalization of the Drude model. Plasmonics 2011; 6(1): 99-104.

Chapter 3

Numerical Simulations of Surface Plasmons Super-Resolution Focusing and Nano-Waveguiding

Xingyu Gao

Additional information is available at the end of the chapter

1. Introduction

Surface plasmons(SPs) are evanescent waves that propagate along the surface of dispersive media [1]. When a light beam incidents onto the surface of metal, the electromagnetic field interacts with the free electrons inside the metal, which leads to the oscillation of free electrons to excite the SPs in the exit medium. Noble metals with nanostructure geometry have special optical properties because they can excite localized surface plasmons (LSPs) under the illumination of light field. Since the pioneer study of Ritchie [2], the special optical properties of SPs in nano-scale have been tightly investigated. Various metallic nano-structures are reported for nano-plasmonic devices, including thin film[3,4], nanowires[5,6], nanorods[7,8], nano-hole array[9,10], nano-slits[11] et. al. Due to the sub-wavelength excitation range and strong field enhancement properties of SPs, they have been widely applied in super-resolution optical microscopy[12], nano-photonic trapping technology[13], biology and medical sciences[14] and nano-photonic waveguide[15].

In this chapter we firstly introduce a numerical algorithm for implementing the relative permittivity model of dispersive media in finite difference time domain(FDTD) method, i.e. piecewise linear recursive revolution(PLRC) method. Next the super-resolution phenomenon derived from the surface plasmons excited at the focal region is presented and analyzed. In the third part, we explore two kinds of plasmonic nano-waveguide: parallel nanorods and metal-dielectric-metal structure. Their optical properties, such as long distance waveguiding from the focal region, turning waveguiding effect and optical switch effect, will be demonstrated in detail. Finally, we conclude our research results in recent years and look forward the applications of surface plasmons in nano-technologies.

2. FDTD method for plasmonics simulation

The FDTD method was firstly proposed by Yee in 1966[16]. Because of its strong and precise power for simulating the propagation of the electromagnetic field, it was quickly applied in many research fields associate with the electromagnetics and optics. The detailed discrete differential equations of Maxwell equations can be found in Ref[17]. In this section, we mainly present the FDTD algorithm for dispersive media which is referred to as piecewise linear recursive revolution(PLRC) method and the verification of the program code.

2.1. 3D FDTD algorithm for dispersive media

The recursive revolution(RC) method for implementing the relative permittivity models of dispersive media in finite difference time domain(FDTD) algorithm was first proposed by Luebbers et al. [18] in 1990. It had been testified to perform faster calculation speed and fewer memory space requirement than the auxiliary differential equation(ADE) method [19], Z-transform method [20] and shift operator(SO) method [21]. In 1996, Kelley and Luebbers proposed the improved PLRC method [22] which remained the speed and low memory of RC method, and provided the accuracy existing in ADE method. The PLRC method for Drude model and Lorentz model has been presented respectively[22, 23]. In this section, we combine these two sets of formulations and propose the PLRC method for Drude-Lorentz model.

The curl equations of Maxwell's equations are presented as:

$$\nabla \times \mathbf{H} = \varepsilon \frac{\partial \mathbf{E}}{\partial t} \tag{1}$$

$$\nabla \times \mathbf{E} = -\mu \frac{\partial \mathbf{H}}{\partial t} \tag{2}$$

where $\mathbf{E}$ is the electric field vector and $\mathbf{H}$ is the magnetic field vector, respectively. ε and μ are the permittivity and permeability of the medium, respectively.

In dispersive medium, the displacement vector $\mathbf{D}$ has the linear relation with the electric field vector $\mathbf{E}$ in frequency domain as:

$$\mathbf{D}(\omega) = \varepsilon(\omega)\mathbf{E}(\omega) = \varepsilon_0\varepsilon_r(\omega)\mathbf{E}(\omega) \tag{3}$$

where $\varepsilon_r(\omega)$ is the relative permittivity function of dispersive medium. It can be expressed by Drude-Lorentz model[24]as:

$$\varepsilon_r(\omega) = \varepsilon_\infty + \frac{\omega_d^2}{j\omega\gamma_d-\omega^2} + \frac{\Delta\varepsilon\omega_L^2}{\omega_L^2+2j\omega\delta_L-\omega^2} = \varepsilon_\infty + \chi_d(\omega) + \chi_L(\omega) \tag{4}$$

where ε_∞ is the relative permittivity at the infinite frequency, ω_d is the plasma frequency, γ_d is the oscillation rate of electrons, $\Delta\varepsilon$ is the difference of the relative permittivity between infinite frequency and zero frequency, ω_L is the oscillation rate of Lorentzian dipoles, δ_L is the decreasing coefficient, $\chi_d(\omega)$ and $\chi_L(\omega)$ are Drude susceptibility term and Lorentz susceptibility term respectively. The update equation for the electric field in FDTD format is:

$$\mathbf{E}^{n+1} = \left(\frac{\varepsilon_\infty - \xi^0}{\varepsilon_\infty - \xi^0 + \chi^0}\right)\mathbf{E}^n + \left(\frac{\Delta t/\varepsilon_\infty}{\varepsilon_\infty - \xi^0 + \chi^0}\right)\nabla \times \mathbf{H} + \left(\frac{1}{\varepsilon_\infty - \xi^0 + \chi^0}\right)\mathbf{\Psi}^n \qquad (5)$$

where $\mathbf{\Psi}^n$ is an intermediary variable which includes the contributions from Drude term and Lorentz term, respectively. The parameters in Equ.(5) are combination of Drude and Lorentz terms. The contribution from Drude term for variable $\mathbf{\Psi}^n$ is updated as:

$$\mathbf{\Psi}_d^n = \left(\Delta\chi_d^n - \Delta\xi_d^n\right)\mathbf{E}^n + \Delta\xi_d^n\mathbf{E}^{n-1} + \mathbf{\Psi}^{n-1}\exp\left(-\gamma_d\Delta t\right) \qquad (6)$$

where

$$\chi_d^n = \frac{\omega_d^2}{\gamma_d}\left[\Delta t - \frac{1}{\gamma_d}\left(1 - e^{-\gamma_d\Delta t}\right)\right] \qquad (7)$$

$$\Delta\chi_d^n = -\frac{\omega_d^2}{\gamma_d^2}\left(1 - e^{-\gamma_d\Delta t}\right)^2 \qquad (8)$$

$$\xi_d^n = \frac{\omega_d^2}{\gamma_d}\left[\frac{\Delta t}{2} - \frac{1}{\gamma_d^2}\left(1 - e^{-\gamma_d\Delta t}\right) + \frac{1}{\gamma_d}e^{-\gamma_d\Delta t}\right] \qquad (9)$$

$$\Delta\xi_d^n = -\frac{\omega_d^2}{\gamma_d^2}\left[\frac{1}{\gamma_d\Delta t}\left(1 - e^{-\gamma_d\Delta t}\right)^2 - \left(1 - e^{-\gamma_d\Delta t}\right)e^{-\gamma_d\Delta t}\right] \qquad (10)$$

The contribution from Lorentz term for variable $\mathbf{\Psi}^n$ is updated as:

$$\widehat{\mathbf{\Psi}}_L^n = \left(\Delta\widehat{\chi}_L^n - \Delta\widehat{\xi}_L^n\right)\mathbf{E}^n + \Delta\widehat{\xi}_L^n\mathbf{E}^{n-1} + \widehat{\mathbf{\Psi}}_L^{n-1}\exp\left((-\alpha + j\beta)\Delta t\right) \qquad (11)$$

where "∧" means it is a complex variable. $\alpha = \delta_L$, $\beta = (\omega_L^2 - \delta_L^2)$, $\gamma = \Delta\varepsilon\omega_L^2$, and

$$\widehat{\chi}_L^n = -\frac{j\gamma}{-\alpha + j\beta}\{1 - \exp[(-\alpha + j\beta)\Delta t]\} \qquad (12)$$

$$\Delta\widehat{\chi}_L^n = \frac{j\gamma}{-\alpha + j\beta}\{1 - \exp[(-\alpha + j\beta)\Delta t]\} \qquad (13)$$

$$\widehat{\xi}_L^n = \frac{j\gamma}{\Delta t(\alpha - j\beta)^2}\{[(\alpha - j\beta)\Delta t + 1]\exp[(-\alpha + j\beta)\Delta t] - 1\} \qquad (14)$$

$$\Delta\widehat{\xi}_L^n = \frac{j\gamma}{\Delta t(\alpha - j\beta)^2}\{[(\alpha - j\beta)\Delta t + 1]\exp[(-\alpha + j\beta)\Delta t] - 1\}\{1 - \exp[(-\alpha + j\beta)\Delta t]\} \qquad (15)$$

Finally the variable $\mathbf{\Psi}^n$ is the sum of Drude term and the real part of Lorentz term:

$$\mathbf{\Psi}^n = \mathbf{\Psi}_d^n + \text{Re}\left(\widehat{\mathbf{\Psi}}_L^n\right) \qquad (16)$$

The parameters in Equ.(5) are also the sum of Drude term and the real part of Lorentz term:

$$\chi^n = \chi_d^n + \text{Re}\left(\widehat{\chi}_L^n\right) \qquad (17)$$

$$\xi^n = \xi_d^n + \text{Re}\left(\widehat{\xi}_L^n\right) \qquad (18)$$

From above equations we can see that the Drude and Lorentz terms are deduced respectively, so that the PLRC method can be applied for both Drude and Drude-Lorentz models. For describing the relative permittivity of silver[25], we adopt the values of the parameters of Drude-Lorentz model as $\varepsilon_\infty=4.6$, $\omega_d=1.401\times10^{16}$Hz, $\gamma_d=4.5371\times10^{13}$Hz, $\Delta\varepsilon=3.428$, $\omega_L=2.144\times10^{16}$Hz, and $\delta_L=1.824\times10^{18}$Hz.

2.2. Verification of the program code

We adopt the classical Krestschmann-Type SPR device shown in Fig.1(a) to verify our FDTD code. The two-dimensional(2D) simulation conditions are: $n_1=1.78$, $n_2=1.132$, $\theta=42°$, thickness of silver film $d=50nm$, and sampling interval $\Delta x=\Delta y=5nm$. Under these configurations, the SPR wavelength calculated by SPs dispersion relation is $\lambda_{SP}=500nm$. A modulated 2D TE Gaussian pulse with the central wavelength $400nm$ is incident from the medium n_1 onto the metal film. The transmission enhancement coefficients in the visible range in the medium n_2 calculated by ADE, SO and PLRC methods are shown in Fig.1(b). These three methods demonstrate the same SPR peak and enhancement strength under the same condition. However, in the shorter wavelength region the ADE method shows larger error, and in the longer wavelength region the SO method shows larger error. So in the total wavelength region, the PRLC method shows better precision than those two methods. We will use the PLRC method and the parameters of Drude-Lorentz model mentioned above throughout this chapter.

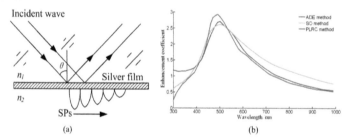

(a)   (b)

Figure 1. Verification of our FDTD code for dispersive media. (a) Configuration of Krestschmann-Type SPR device; (b) Compare of the enhancement coefficients of the three methods.

3. Application of surface plasmons in super-resolution focusing

A tightly focused evanescent field can be generated by a centrally obstructed high numerical aperture objective lens [26], and a super-resolved evanescent focal spot of $\lambda/3$ has been obtained [27]. The enhancement of the electromagnetic (EM) field by tight focusing enables nano-lithography using evanescent field [28]. It has been demonstrated that a tightly focused beam can be further modulated by a negative-refraction layer together with a nonlinear layer [29] or a saturable absorber [30] to approximately $\lambda/4\sim\lambda/5$ close vicinity of the focus. Considering the difficulty in realizing a negative-refraction layer in practice, here we introduce another mechanism of light modulation in the tightly focused region. Due to the use of high numerical aperture objective lens, the focused evanescent field is highly depolarized, which offers strong transverse and longitudinal polarization components.

Therefore the deployment of nano-plasmonic structure, which is polarization sensitive, offers new mechanism to modulate the focused evanescent field. In this section, we will give two research results for the application of surface plasmons in super-resolution focusing.

3.1. Simulation of radially polarized focusing through metallic thin film

As shown in Fig.2, the configuration of tightly focused evanescent field is based on the scanning total internal reflection microscopy[26]. The refractive indices of immersion oil and the coverslip glass are 1.78. The thin silver film is coated on the surface of the coverslip glass. An annular radially polarized incident beam is focused on the upper surface of coverslip glass by an objective. We define the annular coefficient $\varepsilon=d/D$ which produces a ring beam illumination onto the silver film. In the simulation we set $\varepsilon=0.606$ to make sure that the incident angle is larger than the critical angle for the total internal reflection of the ring beam.

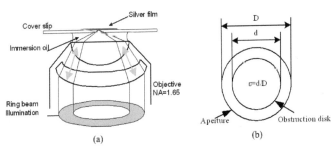

Figure 2. The configuration of focal ring beam illumination for the excitation of SPs. (a) Scheme of total internal reflection focusing; (b) Illustration of annular coefficient.

The electromagnetic field of radially polarized focal beam is expressed as:

$$E(r, \Psi, z) = ik(cos\Psi I_1 \mathbf{i} + sin\Psi I_1 \mathbf{i} + I_0 \mathbf{K}) \tag{19}$$

where

$$I_0 = \int_0^\alpha \sqrt{cos\theta}\ sin^2\theta J_0(krsin\theta)exp(-ikzcos\theta)d\theta \tag{20}$$

$$I_1 = \int_0^\alpha \sqrt{cos\theta}\ sin\theta cos\theta J_1(-krsin\theta)exp(-ikzcos\theta)d\theta \tag{21}$$

where $\mathbf{i}$, $\mathbf{j}$ and $\mathbf{k}$ are unit vectors in the x, y and z directions, respectively. $r = \sqrt{x^2 + y^2}$ is transverse radial coordinate. Ψ denotes the angle between r and $+x$ axis. θ denotes the incident angle of light. $J_i(x)$ is the ith order Bessel function of the first kind. It should be pointed out that the electric fields under the radially polarized focal beam is circular symmetrical. The simulation conditions are set as: incident wavelength $\lambda=532nm$, NA=1.65, sampling interval $\Delta x=\Delta y=\Delta z=5nm$, thickness of silver film $d=60nm$. In this simulation, we calculate the electric field in the transverse plane with the defocal distance $z=-600nm$ as the input source.

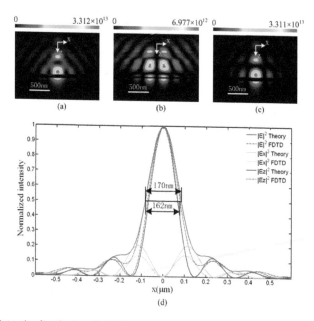

Figure 3. The intensity distribution of radially polarized beam focusing through silver film. Colorbar unit: V^2/m^2. (a) $|E_{total}|^2$; (b) $|E_x|^2$; (c) $|E_z|^2$; (d)Comparison of the sizes of the SPs enhanced focus with that of theoretical normal focus.

Under the conditions mentioned above, after 2000 time steps simulation, the time averaged intensity distributions of E_{total}, E_x and E_z in xz plane are shown in Fig.3. The intensity of total electric field is the combination of E_x and E_z components. The E_z field shows a single evanescent focus which is formed by the SPs excited by the silver film and the total electric field is dominated by E_x component. In order to illustrate the super-resolved SPs focus phenomenon, we plot the SPs focus simulated by FDTD and the theoretical radially polarized focus calculated by Debye theory along x axis in Fig.3(d). The intensity distributions of the SPs focus and the theoretical focus are normalized by the maximum values of their total electric fields. It is demonstrated that the proportion of E_x intensity and Ex intensity of SPs focus is 0.145, while that of the theoretical focus is 0.207, which indicates that the E_z component is enhanced by the SPs. So that the focal size of SPs enhanced focus(FWHM=162nm) is smaller than that of the theoretical focus(FWHM=170nm).

It is obviously that in the total electric field distribution, the reflected field is much stronger than the transmitted field. Under the annular coefficient $0.606<\varepsilon<1.0$ condition, the total ring beam includes a wide range of incident angle($34.18°<\theta<67.97°$). However, only the narrow angle ranges around $\theta=36.72°$ satisfy the SPs dispersion relation. We simulate the thin ring beam illumination ($0.645<\varepsilon<0.65$) which contains the SPR incident angle and the thin ring beam illumination ($0.795<\varepsilon<0.8$) which is far from the SPR incident angle in order to demonstrate the differences between the SPR focus and non-SPR focus.

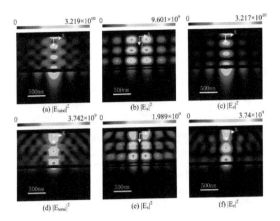

Figure 4. The intensity distributions of SPR focus (a), (b), (c) and non-SPR focus (d), (e), (f) in xz plane. Colorbar unit: V^2/m^2.

The simulation results are shown in Fig.4. Comparing Fig.4(a) and (d), it can be seen that under the SPs dispersion relation, the SPR significantly enhances the focus below the silver film, which leads to a much high transmission rate of the focal field. While for the non-SPR focus, the reflection is much stronger than the transmission, so most part of the focus energy is reflected back. Comparing the E_x components of the two conditions in Fig.4(b) and (e), they demonstrate almost the same distribution with no SPR phenomenon. That is because the E_x component is parallel to the silver film, i.e. s-polarization, it can't excite the SPR. Finally, comparing the E_z components that are perpendicular to the silver film, i.e. p-polarization, of the two conditions in Fig.4(c) and (f), it is obviously demonstrated that the E_z component contributes to the SPR focus dominantly. For the SPR focus, the E_z component is significantly enhanced by the SPR, while for the non-SPR focus, the reflection of E_z component is much larger than its transmission. The SPs dispersion relation satisfied by the ring beam with $0.645 < \varepsilon < 0.65$ is the main mechanism for the super-resolved SPs focus. This phenomenon gives rise to a simple approach for achieving a super-resolution focus beyond diffraction limit, which has potential applications in nano-lithography and nano-trapping.

3.2. Simulation of focusing through two parallel nanorods

3.2.1. Numerical simulation model

Our simulation configuration is shown in Fig.5. Two silver nanorods are lying on the interface of two dielectric media with the separation (D) of $120nm$ centre-to-centre. The refractive index of lower medium (n_1) is 1.78 and the refractive index of upper medium (n_2) is 1.0. The size of the nanorods is $150nm$ in length (l) and $50nm$ in diameter (d). The numerical aperture of the objective is 1.65, and the pure focused evanescent field is generated by inserting a centrally placed obstruction with normalized radius $\varepsilon_c = 0.606$, corresponding to the total internal refraction condition.

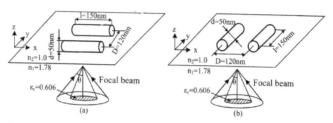

Figure 5. Configuration of nano-plasmonic waveguide. (a) Nanorods lie along x direction. (b) Nanorods lie along y direction.

If the illuminating beam polarizes along the x direction, the electric field at the focal region, calculated with high angle vectorial Debye theory, is highly depolarized, and each polarization component can be expressed as:

$$\mathbf{E}(r, \Psi, z) = \frac{\pi i}{\lambda}\{[I_0 + \cos(2\Psi)I_2]\mathbf{i} + \sin(2\Psi)I_2\mathbf{J} + 2i\cos(2\Psi)I_1\mathbf{K}\} \tag{22}$$

where

$$I_0 = \int_0^\alpha \sqrt{\cos\theta}\sin\theta(1+\cos\theta)J_0(kr\sin\theta)\exp(-ikz\cos\theta)d\theta \tag{23}$$

$$I_1 = \int_0^\alpha \sqrt{\cos\theta}\sin^2\theta J_1(-kr\sin\theta)\exp(-ikz\cos\theta)d\theta \tag{24}$$

$$I_2 = \int_0^\alpha \sqrt{\cos\theta}\sin\theta(1-\cos\theta)J_2(kr\sin\theta)\exp(-ikz\cos\theta)d\theta \tag{25}$$

where $\mathbf{i}$, $\mathbf{j}$ and $\mathbf{k}$ are unit vectors in the x, y and z directions, respectively. $r = \sqrt{x^2+y^2}$ is transverse radial coordinate. Ψ denotes the angle between r and $+x$ axis. θ denotes the incident angle of light. $J_i(x)$ is the ith order Bessel function of the first kind. The electric and magnetic fields are calculated at a plane one wavelength before the interface as the input source for the FDTD simulation. The total field/scatter field technique is used to eliminate the light propagating to $-z$ direction, so that the incident focal beam only propagates in forward direction.

In the simulation, the wavelength of incident focal beam is $532nm$. It should be noted that the annular beam illumination includes a wide range of incident angle ($34.18°<\theta<69.97°$), which corresponds to a wide range of surface plasmon resonance (SPR) wavelengths ($340nm<\lambda<1120nm$). Our simulations show that different wavelengths selected in this spectrum range make little difference for the excitation of LSPs by the nano-plasmonic waveguide. The grid sizes Δx, Δy and Δz for each dimension are set to $2.34nm$. According to the discretization in space domain, the discretization in time domain that satisfies the Courant stability condition is adopted as:

$$\Delta t = \frac{0.985}{c\sqrt{1/\Delta x^2 + 1/\Delta y^2 + 1/\Delta z^2}} \tag{26}$$

where c is the velocity of light in vacuum. Under these configurations, the iteration of program runs 400 time steps for a period. In order to obtain the stable state of the optical field, 10 periods are adopted, i.e. 4000 time steps are calculated in the simulation.

3.2.2. Simulation results and discussion

The intensity distribution of a focused evanescent field under the linearly polarized illumination is shown in Fig.6 and agrees well with previous theory[26, 27]. Under the conditions described above, the intensity of E_y component is one order smaller than either E_x or E_z, so the overall impact of E_y on the intensity distribution is less significant. Due to the depolarization effect, a strong longitudinal E_z component appears, with its strength comparable to the illuminating polarization component E_x, i.e., $|E_z|^2/|E_x|^2\approx0.85$. As a result, the intensity distribution of the total field is splitted to two lobes shown in Fig.6 (a).

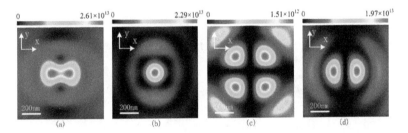

Figure 6. The intensity distribution of evanescent electric field for linearly polarized focal beam at the interface. (a) $I=|E_x|^2+|E_y|^2+|E_z|^2$ (b) $|E_x|^2$ (c) $|E_y|^2$ (d) $|E_z|^2$ (Unit: V^2/m^2)

The modulation of the focused evanescent field by a pair of silver nanorods is demonstrated in Fig. 7, where the intensity distributions in planes of different distances from the interface are illustrated. In the left column, when there is no nanorod on the interface, the focal spot splits into two lobes at different distances above the interface. In the middle column, the nanorods are lying along x direction, parallel to the dominant transverse polarization component E_x. In the right column, the nanorods are lying along y axis, perpendicular to the dominant transverse polarization component E_x. In Fig.7 (a), it is noted that at the xy plane $25nm$ above the interface, the electric field is significantly enhanced by the LSPs between the nanorods, if the nanorods are lying in the y direction, resulting in a strong localized field between the two nanorods. While the nanorods are lying in the x direction, the electric field shows less enhancement and localization. At the plane $60nm$ above the interface ($10nm$ above the nanorods, Fig.7 (b)), it is observed that the longitudinal electric field component E_z displays a strong enhancement at both ends of each nanorod, forming four strong intensity lobes. When the nanorods are lying along x direction, the dominant transverse electric field component E_x is not significantly enhanced, so the electric field is dominated by the longitudinal component E_z, which shows four strong intensity lobes. However when the nanorods lie in the y direction, due to the significant enhancement of E_x component between the nanorods, the four strong lobes become less evident. At the horizontal planes that further away from the nanorods, i.e. at the planes $75nm$ (Fig.7 (c)) and $100nm$ (Fig.7 (d)) above the interface, super-resolved focal spots are demonstrated. In particular, with the significant enhancement of dominant transverse electric field component E_x, the focal spots show narrower distribution and stronger strength, when the nanorods lie in the direction perpendicular to the E_x component.

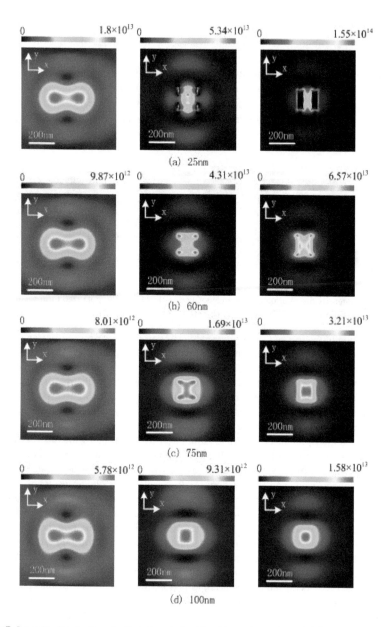

(a) 25nm

(b) 60nm

(c) 75nm

(d) 100nm

Figure 7. Intensity distributions for linearly polarized focal beam in xy plane of different distances from the interface. In the left column there is no nanorod. In the middle column the nanorods are lying along x direction. In the right column the nanorods are lying along y direction. (Unit: V^2/m^2)

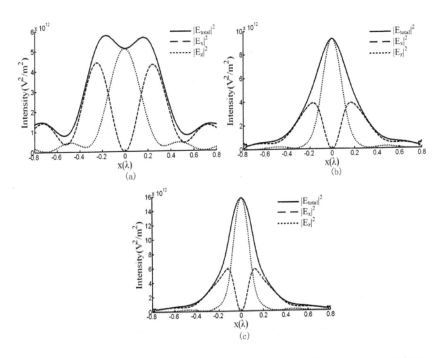

Figure 8. Cross sections for polarization components of linearly polarized light along x axis at the xy plane 100nm above the interface. (a) Without nanorods; (b) Nanorods lie along x direction; (c) Nanorods lie along y direction.

The detailed analysis of the LSPs effect on each polarization component at the plane $100nm$ above the interface is illustrated in Fig.8. When the nanorods are lying along x direction (Fig. 8(b)), the enhancement of E_z component is insignificant compared with the case that the nanorods lie in the y direction. Nevertheless, it still produces a narrower focal spot with 46.6% reduction in FWHM and 61% increase in strength. When the nanorods are lying in y direction (Fig. 4(c)), the dominant transverse polarization E_z is not only significantly enhanced in strength, e.g. by a factor of 2.71, but also becomes more localized, e.g. the full width at half maximum (FWHM) is reduced from $388nm$ to $149nm$. This phenomenon indicates the LSPs excited by the E_z component couple between the nanorods and re-distribute the energy of electromagnetic field. The longitudinal component E_z is also enhanced by a factor of 1.4, and becomes narrower, e.g. the distances between two intensity peaks reduced from $266nm$ to $127nm$. As a result, a super-resolution focal spot can be formed outside the waveguide. The above analysis shows that the nano-plasmonic waveguide provides strongest enhancement to the transverse polarization component perpendicular to the nanorods, followed by the longitudinal polarization component, and the enhancement for the transverse polarization component parallel to the nanorods is least significant.

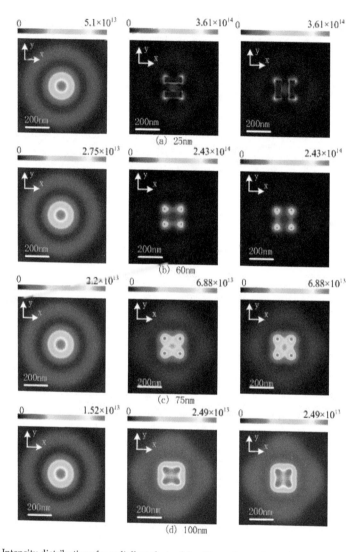

Figure 9. Intensity distributions for radially polarized focal beam in xy plane of different distances from the interface. In the left column there is no nanorod. In the middle column the nanorods are lying along x direction. In the right column the nanorods are lying along y direction. (Unit: V^2/m^2)

It is well known that the focal spot for radially polarized beam is circularly symmetrical. According to the analysis demonstrated in the previous section, it is expected that the circular symmetry would be broken due to the LSPs effect which is polarization sensitive. Fig.9 shows the intensity distributions of evanescent radially polarized focal beam at xy

planes of different distances above the interface for three cases, including without nanorods, nanorods lying in the x and y directions, respectively. The focal spot is circularly symmetrical and decay exponentially further from the interface without the presence of the nanorods, as shown in the left column of Fig.9. With strong LSPs effect from the nano-plasmonic waveguide, the circular symmetry is broken, and the focal spot shows a strong intensity lobe at each of the four ends of the two nanorods (Figs. 9 (b-d)). It is noted that the intensity of the longitudinal component E_z is approximately one order of magnitude stronger than that of the transverse components. As a result, the intensity distribution at the focal region shows four strong intensity lobes produced by LSPs effect excited by the longitudinal polarization component E_z.

In this section we demonstrate a new method to modulate highly focused evanescent field with a nano-plasmonic waveguide. The modulation of focus is based on the mechanism that the LSPs are polarization sensitive and the focus is strongly depolarized by a high numerical aperture objective. For a simple nano-plasmonic waveguide that consists of two silver nanorods lying on the interface between two dielectrics, LSPs effect is strongest for the polarization component perpendicular to the nano-plasmonic waveguide. A super-resolved focal spot with significantly enhanced strength can be achieved, when the nanorods are lying perpendicular to the dominant polarization component. The design of the nano-plasmonic waveguide structure gives rise to a new approach to further improve the tightly focused evanescent field to achieve the resolution beyond diffraction limit, and thus facilitates potential applications in nano-trapping and nano-lithography.

4. Plasmonic nano-waveguiding

In this section we will propose the waveguiding properties of two types of plasmonic nano-structures: parallel long nano-wires and the metal-dielectric-metal waveguide. They all present interesting optical nano-waveguiding properties and can be applied in some special research areas.

4.1. Focusing through parallel nanorods that perpendicular to the interface

In recent years, metallic nanorods are intensely investigated and widely applied in the field of super-long waveguiding [31]. It has been reported that the optical properties of single nanorod is sensitive to the polarization of incident wave and the rod aspect ratio [32]. While for the two parallel nanorods and U-shaped nanorods, their optical properties are more complicated and interesting [33]. The SPPs excited along the surface of single nanorod couple between the nanorods, which leads to the extinction spectrum whose resonance peaks are dependent on the geometry of the nanorods. The U-shaped nanorods show stronger resonance strength and more longitudinal couple modes than the two parallel nanorods.

In some cases, such as super-resolution imaging or focusing, the more complicated focusing beam with three polarization components should be considered as the incident source.

However, the interaction between the focal beam and the two nanorods has not been investigated thoroughly yet. Here we fist present the SPR excitation spectrums of two parallel silver nanorods and the π-shaped silver nanorods using the FDTD method. Second we simulate the focusing process through these two structures using 3D FDTD method, respectively. The waveguide effect and electromagnetic field transfer efficiency of these two structures are compared and analyzed. Finally, the focusing process through the angular π-shaped nanorods structure is simulated which presents the ability of guiding the focal field to different directions.

4.1.1. Simulation modeling and resonance spectrums

The schemes of two silver nanorods structures are shown in Fig.10. In Fig.10(a), the length of nanorods is l, the diameter of nanorods is a and the distance between the two nanorods is d. In Fig.10(b), the length of longitudinal nanorods is l_1, the length of transverse nanorod is l_2. In our simulations, we set $a=50nm$, $l=l_1=500nm$, and $l_2=260nm$ constantly, and change the distance d from 30nm to 100nm gradually to calculate their SPR spectrums.

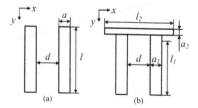

Figure 10. Structures of silver nanorods waveguide. (a) Two parallel nanorods structure; (b) π-shaped nanorods structure

An x-polarized Gaussian temporal profile pulse with the central wavelength of 400nm is injected from the upper side and propagates through the nano-structures. The spectrum width of the pulse is wide enough to cover the wavelength range from 300nm to 1000nm. The temporal electromagnetic field coupling between the nanorods at the central cross section of the nanorods is recorded, and transformed to frequency domain using Fourier transform, and divided by the incident wave magnitude at each wavelength component to calculate the enhancement coefficient spectrum.

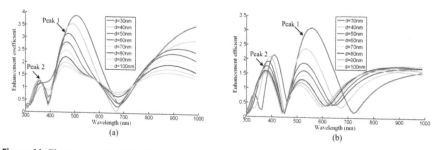

Figure 11. Electromagnetic field enhancement coefficient of the SPR coupling between the two silver nanorods. (a) Two parallel nanorods structure; (b) π-shaped nanorods structure.

As shown in Fig.11, the spectrums perform two peaks for both structures, but the positions of the peaks of the two structures under the same distance d are different. In Fig.11(a), for the two parallel nanorods structure, the main resonance peak 1 shows blue shift with the increase of the distance d, while the sub-resonance peak 2 remains at the wavelength about $350nm$. In Fig. 11(b), the main resonance peak 1 also shows blue shift with the increase of the distance d. The peak 2 remains at about $390nm$ but becomes stronger. Due to the reflection by the transverse nanorod in the π-shaped nanorods structure, their enhancement coefficients are a little smaller than those of the two parallel nanorods structure.

4.1.2. Focusing through nanorods structures

Based on the results obtained above, we will investigate the focusing processes through the nanorods structures in this section. Due to the depolarization effect of high numerical aperture objective, the incident linearly polarized beam would change its polarization state after propagating through the objective, i.e. the strong longitudinal polarization component occurs. The interaction effect between the incident focal beam and the two nanorods structures are more complicated than that of the single polarized incident beam case. Here the focal beam is calculated by the vectorial Debye theory and induced into the 3D FDTD simulation region with total-field/scatter-field method. The refractive index of upper medium is 1.78 and the refractive index of lower medium is 1.0. The numerical aperture of the objective is 1.65, and the pure focused evanescent field is generated by inserting a centrally placed obstruction disk with the normalized radius $\varepsilon_c=0.606$, corresponding to the total internal refraction condition at the interface between the two media. We fix the distance $d=70nm$ throughout the following simulations, so that the tops of the two parallel nanorods are all covered in the focal region. The wavelength of focal beam is $532nm$ which corresponding to the resonance peak of π-shaped nanorods structure at $d=70nm$ condition. We place the nanorods structures normally onto the interface, and set the length of the longitudinal parallel nanorods to $1250nm$ to observe the long waveguide effect for transferring the electromagnetic field from the focus region to the far field. The 3D discretized cell sizes for three directions are $\Delta x=\Delta y=\Delta z=4.68nm$ and the time step $\Delta t=8.27288e-18s$ which satisfies the Courant stability condition. After 4000 time steps simulation, the total electromagnetic field is stable and the time averaged intensity distributions are obtained as shown in Fig.12.

The long waveguide effect produced by the nanorods structures are clearly shown in Fig.12. Without the nano-structures, the pure evanescent focal field would decay exponentially away from the interface and only can propagate no more than $100nm$. However, when the nanorods structures are implemented onto the interface, the SPPs are excited by the evanescent focal field along the surfaces of the nanorods. The electromagnetic fields along the inner surface of the longitudinal nanorods are stronger than those along the outer surface of the longitudinal nanorods, which means that the SPPs between the two nanorods couple with each other and form a long waveguide. The SPPs propagate along the nanorods with a resonance mode, with each space resonance period about $150nm$ in length. The electromagnetic field can be transferred to the lower end of the nanorods where is the far field more than $1\mu m$.

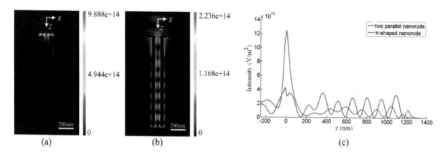

Figure 12. Focusing through long nanostructure waveguide. (a) Two parallel nanorods structure; (b) π-shaped nanorods structure; (c) Plot of intensity along the z axis. Color bar unit: V²/m².

The two parallel nanorods structure and the π-shaped nanorods structure show almost the same SPPs resonance mode, except their excitation strength. In Fig.12(a), the electromagnetic field is strongly enhanced at the top of the two nanorods and decay exponentially very fast, which leads to four strong field points at the top of the nanorods. In Fig.12(b), the focal field is evenly enhanced by the transverse nanorod, so there is no strong field point shown at the top side. The comparison of the electromagnetic field intensities along z axis of the two situations is shown in Fig.12(c). At the interface of the two media(0 position along z axis), the two parallel nanorods show stronger intensity than the π-shaped nanorods. However, at the distances larger than 200nm along z axis, the π-shaped nanorods show stronger intensity than the two parallel nanorods. There are two reasons for this phenomenon. First, under $d=70nm$ condition, the incident wavelength 532nm is almost the SPR wavelength for the π-shaped nanorods structure as shown in Fig.11(b), but not the SPR wavelength for the two parallel nanorods structure as shown in Fig.11(a). Second, the transverse nanorod of π-shaped nanorods interacts with the longitudinal polarization component of the focal beam, so that the longitudinal SPPs along the surfaces of the nanorods are excited, which gives contribution to the total SPPs intensity in the nanorods waveguide. The π-shaped nanorods structure shows stronger electromagnetic field enhancement and better energy transfer efficiency in the far field than the two parallel nanorods structure. This result gives a method to transfer the focal field from the focus region to the far field.

Finally, we present the simulation of focusing through the angular π-shaped nanorods structure as shown in Fig.13(a). The two legs of the π-shaped nanorods are bent by the angle of 90°, so that the ends of the nanorods shift from the optical axis with the distance of l_2. In our simulation, we set $l_1=500nm$, $l_2=500nm$, $l_3=260nm$, $d=70nm$ and $a=50nm$. The incident wavelength keeps 532nm. The simulation results for different cross section planes are shown in Fig.13(b), (c) and (d). In Fig.13(b), the intensity distribution in the xz plane shows the same pattern with that in Fig.12(b). In Fig.13(c), the intensity distribution in the yz plane shows an "L" shaped optical path which formed by the SPPs coupled between the two nanorods. This phenomenon testifies that the SPPs can propagate not only along the straight path, but also along the angular path with the angular structure of the nanorods. The focal electromagnetic field can be transferred not only along the optical axis, but also to the directions that perpendicular to the optical axis. In Fig.13(d), the xy plane is the cross section through the centers of the lower two transverse nanorods. It shows the same resonance length with the longitudinal part of the waveguide. It can be predicted that if we change the bend angle to other degrees, the SPPs also can be transferred to other directions. Due to the limit memory of computer, we can't simulate the waveguiding effect of longer nanorods structures, but it can be seen that the electromagnetic field waveguiding ability of the two nanorods structures investigated in this paper must be far more than $1\mu m$.

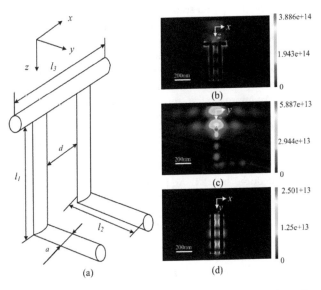

Figure 13. Focusing through angular π-shaped nanorods structure. (a) The configuration of the angular π-shaped nanorods structure; (b) The intensity distribution in xz plane; (c) The intensity distribution in yz plane.

In this section, the optical properties of two silver nanorods plasmonic waveguide structures are simulated with the FDTD method. The SPR spectrums of the two parallel nanorods and the π-shaped nanorods structures are calculated, which show different SPR resonance peaks at the same distance condition for both structures. The focusing processes through these two types of structures show almost the same SPPs resonance mode with each resonance period about 150nm. However, at the far field, the electromagnetic field enhancement of the π-shaped nanorods structure is stronger than that of the two parallel nanorods, due to the contribution of longitudinal SPPs component excited by the transverse nanorod. At last, the focusing process through the angular π-shaped nanorods structure indicates that the focus field can be transferred not only to the far field along the optical axis, but also to any other directions that the ends of the two nanorods direct to. The two nanorods plasmonic waveguide structures studied in this paper have potential applications in nano-biosensing, nano-trapping or nano-waveguiding.

4.2. MDM nano-waveguides

Plasmonic waveguides based on the principles of SPPs have gained great attentions in recent years due to their ability of confining and guiding optical field in sub-wavelength scale. Among various types of plasmonic waveguides, the metal-dielectric-metal (MDM) waveguide is considered to be a key element in the fields of waveguide couplers[34, 35], sub-wavelength scale light confinement[36, 37], wavelength filters[38] and integrated optical devices[39, 40]. The remarkable advantages of the MDM waveguide, including the strong confinement of optical field in nano-scale gaps, the high sensitivity of its transmission characteristics to the waveguide structures, and the facility of its fabrication, attracted a great deal of effort to be devoted to develop the MDM based nano-plasmonic devices. It has been reported that the transmission of MDM waveguide coupled with stub structure could be changed with the length of the stub[38]. Further more, the stub filled with absorptive medium was considered to be a resonance cavity that acts as an optical switch controlled by the pumping field[39]. An improved transmission model[40] was also developed to describe the transmittance of multi-stubs MDM structure.

The investigation approaches for the transmission characteristics of MDM waveguide include theoretical transmission line theory(TLT) and FDTD method [40]. In this section, we first provide the formulas of the optical transmission characteristics of the MDM waveguide with a stub structure deduced by TLT. And then the FDTD method will be employed to numerically study the optical switch effect of the stub structure in terms of changing the length and the refractive index of the stub, respectively. The simulation results coincide with the calculation results of TLT and the physical mechanisms of the optical switch effect are analyzed and discussed.

4.2.1. Scheme modeling and transmission line theory

The scheme of the MDM waveguide coupled with a single stub structure is shown in Fig.14(a). Firstly we assume that the medium in the stub is air, i.e. $n_1=n_2=1.0$. The silver is employed as the dispersive medium for the MDM waveguide. The width of the waveguide and the stub is $d=100nm$. A transverse electric (TE) plane wave with a wavelength $\lambda=600nm$ is incident from the left side into the waveguide. The transmission line modeling of the schematic is shown in Fig.14(b).

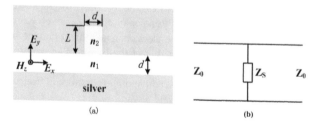

(a)　　　　　　　　　　　　(b)

Figure 14. (a) The scheme of MDM waveguide coupled with a stub structure with the same width d and the length L. (b) The transmission line modeling of the scheme.

According to the transmission line theory, the stub can be considered as admittance. Assume that the phase shift only occurs when the SPPs are reflected by the end of the stub. Z_0 and Z_s are the characteristic impedances of the loss-free transmission lines corresponding to the MDM waveguide and the stub[40], respectively. Their relation is expressed as:

$$\frac{1}{Z_s} = -j\frac{1}{Z_0}\tan\left(\frac{2\pi L}{\lambda_{SP}}\right)$$

(27)

where L is the length of the stub and λ_{SP} is the propagating wavelength of the SPPs. From Fig.14(b), the amplitude transmission of the electric field can be expressed as:

$$t = \frac{2Y_0}{2Y_0 + Y_s}$$

(28)

where $Y_0 = 1/Z_0$ and $Y_s = 1/Z_s$. Therefore, the energy transmission of the MDM waveguide coupled with a single stub is finally expressed as[38]:

$$T = \frac{4}{4 + \tan^2\left(\frac{2\pi L}{\lambda_{SP}}\right)}$$

(29)

From Equ.(29) we can see that the energy transmission is the function of L and λ_{SP}. In the stub, the SPPs wavelength λ_{SPS} can be changed by the refractive index of the medium in the stub. As a result, the transmission would be modulated periodically by changing the length L and the refractive index n_2 of the stub linearly.

4.2.2. Numerical simulation modeling and results

We simulated the transmission of the MDM waveguide with L ranging from 0 to $1\mu m$ by 2D FDTD algorithm. The relative permittivity of silver is described by Drude-Lorentz model to fit the experimental data of relative permittivity of silver and implemented into FDTD program with the PLRC method. The discretized cell size is $\Delta x = \Delta y = 5nm$. After 3000 time steps simulation, the results are shown in Fig.15.

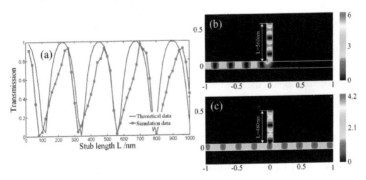

Figure 15. The transmission of MDM waveguide coupled with a stub structure as a function of the stub length. (a) Compare of the transmission calculated by transmission line theory and FDTD simulation. (b) The intensity distribution of the MDM waveguide at the "off" state with $L=560nm$. (c) The intensity distribution of the MDM waveguide at the "on" state with $L=480nm$.

From Fig.15(a) we can see that the simulation results agree well with the theoretical data calculated by Equ.(29). The SPPs excited along the surfaces of the metal layers propagate in the waveguide with a resonance mode. When passing through the stub, a part of the SPPs propagate into the side-coupled stub. The SPPs reflected from the end of the stub interfere with the passing SPPs, which leads to a modulation of the superposition wave. If we change the length of the stub, the phase change of the interference would cause the transmission vary from 0 to 1, so that the MDM waveguide coupled with a single stub structure presents the optical switch effect on the incident wave. As the incident wavelength λ=600nm, the λ_{SP} is approximately to be 480nm. The transmission shows about 4.5 periods for L linearly ranging from 0 to 1μm in length. The lowest transmissions are less than 1%, and the highest transmissions are more than 92%, which means the transmission of this structure switches between "on" and "off" states with the variation of the length of the stub. As shown in Fig.15(b), when the transmission is at the "off" state with L=560nm, the reflected light from the end of the stub modulates the interference intensity to be 0, which causes the propagating SPPs in the waveguide stopped by the stub significantly. As shown in Fig. 15(c), when the transmission is at the "on" state with L= 480nm, the SPPs can be transferred to the end of the waveguide without any loss. For both states, the superposition of the waves going into the stub and reflected by the end of the stub makes the field intensity in the stub stronger than that in the waveguide. Moreover, the field intensity of "off" state is stronger than that of "on" state due to the storage of the stopped field energy in the stub by the interference.

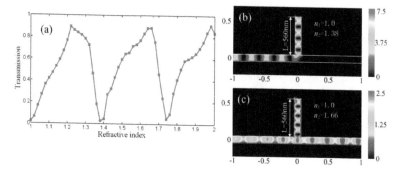

Figure 16. (a)The transmission of the MDM waveguide coupled with a stub structure as a function of the refractive index n_2 of the stub. (b) The "off" state when n_2=1.38. (c) The "on" state when n_2=1.66.

Another approach for modulating the phase of the reflected wave from the end of the stub is changing the refractive index of the stub. We carried out a series of simulations with the refractive index n_2 ranging from 1.0 to 2.0, and keeping n_1=1.0. The simulation results are shown in Fig.16. In Fig.16(a), it is obviously that the transmission demonstrate a periodic distribution as a function of the refractive index n_2. This phenomenon is similar with the distribution as shown in Fig.15(a). The mismatch of n_1 and n_2 leads to the phase difference between the reflected wave in the stub and the passing wave in the waveguide which also processes a phase modulation effect for the interference wave. Fig.16(b) and (c) show the phenomena of "off" and "on" states of the transmission when n_2=1.38 and n_2=1.66, respectively. It is clearly shown that the effective wavelength of the propagating SPPs in the stub is changed according to the refractive index n_2. The field intensity at "off" state is also much higher than that at "on" state because of the storage of the SPPs energy in the stub.

We have numerically studied the transmission characteristics of the MDM waveguide coupled with stub structure as functions of the length and the refractive index of the stub, respectively. The 2D FDTD simulation results show that the transmission rates obtained by both approaches change as periodical distributions, which implies that the MDM waveguide can be treated as an optical switch device controlled by the length and the refractive index of the stub. The physical mechanism of this phenomenon is the phase modulation of the interference of the reflected SPPs wave from the end of the stub and the passing SPPs wave in the waveguide. The results help us to further apply the MDM waveguide as an optical switch element in nano-scale optical chips and optical integrated devices.

5. Conclusions

The importance of surface plasmons in the applications of nano-photonics has been proved in many examples. In this chapter we focused on two fields: super-resolution focusing and nano-waveguiding. Super-resolution focusing is the key element for nano-lithography, high density optical data storage, and super-resolution imaging. We have presented that the high density focal field can be re-distributed by some specified metallic nano-structures such as nano-film and parallel nanorods, so that a super-resolution focusing can be generated in some particular space areas. This method can be further applied in optical nano-trapping due to the small and enhanced plasmonic focus. While for nano-waveguiding there have been many nano-structures and methods reported previously. Here we have presented the parallel nanorods and MDM structures, respectively. When the parallel nanorods are perpendicularly put on the interface of two materials, the incident focal field generates surface plasmons along their surfaces which form a nano-waveguide. The fields of focus can propagate along the waveguide for a long distance. The MDM with a stub structure showed a optical switch effect with the altering of the length and the refractive index of the stub, which has a potential application in optical communications and optical sensing.

Author details

Xingyu Gao

Institute of Opto-mechatronics, Guangxi Key Laboratory of Manufacturing System & Advanced Manufacturing Technology, School of Mechanical & Electrical Engineering, Guilin University of Electronic Technology, Guilin Guangxi, China

6. References

[1] William L.Barnes, Alain Dereux and Thomas W. Ebbesen. Surface plasmon subwavelength optics. Nature 2003; 424(6950) 824-830.

[2] R.H.Ritchie. Plasma Losses by Fast Electrons in Thin Film. Phys. Rev. 1957; 106(5) 874-881.

[3] Hiroshi Kano, Seiji Mizuguchi, and Satoshi Kawata, Excitation of surface-plasmon polaritons by a focused laser beam, J. Opt. Soc. Am. B 1998; 15(4) 1381-1386.

[4] Lihong Shi and Lei Gao. Subwavelength imaging from a multilayered structure containing interleaved nonspherical metal-dielectric composites. Phys. Rev. B 2008; 77(19) 195121.

[5] Thierry Laroche and Christian Girard. Near-field optical properties of single plasmonic nanowire. Appl. Phy. Lett. 2006; 89(23) 233119.

[6] Stephen K. Gray and Teobald Kupka. Propagation of light in metallic nanowire arrays: Finite-difference time-domain studies of silver cylinders. Phy. Rev. B 2003; 68(4) 045415.

[7] Yuan-Fong Chao, Min Wei Chen and Din Ping Tsai. Three-dimensional analysis of surface plasmon resonance models on a gold nanorod. Appl. Opt. 2009; 48(3) 617-622.

[8] Xingyu Gao and Xiaosong Gan. Modulation of evanescent focus by localized surface plasmons waveguide. Opt. Express 2009; 17(25) 22726-22734.

[9] Fu Min Huang, Nikolay Zheludev, Yifang Chen, et al.. Focusing of light by a nanohole array. Appl. Phy. Lett. 2007; 90(9) 091119.

[10] Yakov M. Strelniker. Theory of optical transmission through elliptical nanohole arrays. Phy. Rev. B 2007; 76(8) 085409.

[11] Haofei Shi, Changtao Wang, Chunlei Du, et al. Beam manipulating by metallic nano-slits with variant widths. Opt. Express 2005; 13(18) 6815-6820.

[12] Michael G. Somekh, Shugang Liu, Tzvetan S. Velinov, et al. High-resolution scanning surface-plasmon microscopy. Appl. Opt. 2000; 39(34) 6279-6287.

[13] Maurizio Righini, Giovanni Volpe, Christian Girard, et al. Surface Plasmon Optical Tweezers: Tunable Optical Manipulation in the Femtonewton Range. Phy. Rew. Lett. 2008; 100(18) 186804.

[14] Patrick Englebienne, Anne Van Hoonacker and Michel Verhas. Surface plasmon resonance: principles, methods and applications in biomedical sciences. Spectroscopy, 2003; 17: 255-273.

[15] Jason M. Montgomery and Stephen K. Gray. Enhancing surface plasmon polariton propagation lengths via coupling to asymmetric waveguide structures. Phys. Rev. B 2008; 77(12) 125407.

[16] K.S.Yee. Numerical Solution of Initial Boundary Value Problems Involving Maxiwell's Equations in Isotropic Media. IEEE Trans. Antennas Propagat. 1966; 14(3) 802-807.

[17] A. Taflove and S. C. Hagness. Computational Electrodynamics: The Finite-Difference Time-Domain Method. 3rd ed, Norwood, MA: Artech House; 2005.

[18] Raymond J. Luebbers, Forrest P. Hunsberger, Karl S. Kunz, et al. A Frequency-Dependent Finite-Difference Time-Domain Formulation for Dispersive Materials. IEEE Trans. Electromag. Compat. 1990; 32(3) 222-227.

[19] Rose M. Joseph, Susan C. Hagness, and Allen Taflove. Direct time integration of Maxwell's equations in linear dispersive media with absorption for scattering and propagation of femtosecond electromagnetic pulses. Opt. Lett. 1991; 16(8) 1412-1414.

[20] Dennis M. Sullivan. Frequency-Dependent FDTD Methods Using Z Transforms. IEEE Trans. Antennas Propag. 1992; 40(10) 1223-1230.

[21] Ge Debiao, Wu Yueli, Zhu Xiangqin. Shift operator method applied for dispersive medium. Chinese J. Radio Sci. 2003; 18(4) 359-362.

[22] David F. Kelley and Raymond J. Luebbers. Piecewise Linear Recursive Convolution for Dispersive Media Using FDTD. IEEE Trans. Antennas Propag. 1996; 44(6) 792-797.

[23] Jun SHibayama, Taichi Takeuchi, Naoki Goto, et al. Numerical Investigation of a Kretschmann-Type Surface Plasmon Resonance Waveguide Sensor. J. Lightw. Technol. 2007; 25(9) 2605-2611.

[24] Xingyu Gao, Zexin Xiao, Lihua Ning. Surface plasmon enhanced super-resolution focusing of radially polarized beam. OSA-IEEE Topical Conference: Advances in Optoelectronics & Micro/nano-Optics. 2010:5713551.

[25] P.B. Johnson and R. W. Christy. Optical Constants of the Noble Metals. Phys. Rew. B 1972; 6(12) 4370-4397.

[26] James W. M. Chon and Min Gu. Scanning total internal reflection fluorescence microscopy under one-photon and two-photon excitation: image formation. Appl. Opt. 2004; 43(5) 1063-1071.

[27] Baohua Jia, Xiaosong Gan, and Min Gu. "Direct measurement of a radially polarized focused evanescent field facilitated by a single LCD," Opt. Express 13, 6821-6827.

[28] Baohua Jia, Xiaosong Gan, and Min Gu. Direct measurement of a radially polarized focused evanescent field facilitated by a single LCD. Opt. Express 2005; 13(18) 6821-6827.

[29] A. Husakou and J. Herrmann. Subdiffraction focusing of scanning beams by a negative-refraction layer combined with a nonlinear layer. Opt. Express 2006; 14(23) 11194-11203.

[30] A. Husakou and J. Herrmann. Focusing of Scanning Light Beams below the Diffraction Limit without Near-Field Spatial Control Using a Saturable Absorber and a Negative-Refraction Material. Phy. Rev. Lett. 2006; 96(1) 013902.

[31] R. F. Oulton, V. J. Sorger, D. A. Genov, et al. A hybrid plasmonc waveguide for subwavelength confinement and long-range propagation. Nature Photonics 2008; 2(8) 496-500.

[32] Yuan-Fong Chau, Din Ping Tsai, Guang-Wei Hu, et al. Subwavelength optical imaging throught a silver nanorod. Opt. Eng., 2007; 46(3) 039701.

[33] Zhongyue Zhang and Yiping Zhao. Optical properties of U-shaped Ag nanostructures. J. Phy: Condens. Matter 2008; 20(34) 345223.

[34] G. Veronis and S. Fan. Subwavelength light bending by metal slit strctures. Opt. Express 2005; 13(24) 9652-9659.

[35] R. A. Wahsheh, Z. Lu and M. A. G. Abushagur. Nanoplasmonic couplers and splitters. Opt. Express 2009; 17(21) 19033-19040.

[36] Ki Young Kim, Young Ki Cho, and Heung-Sik Tae. Light transmission along dispersive plasmonic gap and its subwavelength guidance characteristics. Opt. Express 2006; 14(1) 320-330.

[37] Y. C. Jun, R. D. Kekatpure, J. S. White, and M. L. Brongersma. Nonresonant enhancement of spontaneous emission in metal-dielectric-metal plasmon waveguide structures. Phy. Rev. B 2008; 78(17) 153111.

[38] Changjun Min and Georgios Veronis. Absorption switches in metal-dielectric-metal plasmonic waveguides. Opt. Express 2009; 17(13) 10757-10766.

[39] A. Pannipitiya, I. D. Rukhlenko, M. Premaratne, H. T. Hattori, and G. P. Agrawal. Improved transmission model for metal-dielectric-metal plasmonic waveguides with stub structure. Opt. Express 2010; 18(6) 6191-6204.

[40] Georgios Veronis, Zongfu Yu, Şükrü Ekin Kocabaş, David A. B. Miller, Mark L. Brongersma, and Shanhui Fan. Metal-dielectric-metal plasmonic waveguide devices for manipulating light at the nanoscale. Chin. Opt. Lett. 2009; 7(4) 302-308.

Plasmonic Structures for Light Transmission, Focusing and Guiding

Chapter 4

Surface Plasmons on Complex Symmetry Nanostructured Arrays

Brian Ashall and Dominic Zerulla

Additional information is available at the end of the chapter

1. Introduction

In recent years it has become accepted that the direction of the plasmonics community is becoming increasingly applied. This is a natural progression, whereby scientific advances are inevitably applied to appropriate technologies. Indeed, in order for the community of plasmonics to continue growing, or at least to maintain the current status, real world technological applications are required. However, this is not to say that the level of fundamental SP research will decrease, as there are still many questions to be answered or clarified on a fundamental level. The dramatic growth of plasmonics in the modern era can be predominantly contributed to four components: nanoscale fabrication techniques, computation power, SPP applications, and "the promise of plasmonics" [1].

The focus of this chapter draws inspiration from all four of the above points. In particular, following an introduction and description of nanostructure fabrication techniques and design considerations in the first section, the second section will detail farfield analysis techniques used for the examination of the light diffracted from structured arrays, and the subsequent identification of plasmons based on their farfield signatures. Following this, the excitation of SPPs on tailor designed 3 fold symmetric structures will be discussed, with advantages resulting from this symmetry breaking explored. Unlike rotationally symmetric structures, such 3-fold symmetric structures are inherently capable of symmetry breaking as a result of their orientational dependencies. In particular, this section (Sec. 3) will focus on the engineering of the SPP nearfield distributions on complex nanostructures. For this, the use of a PEEM (Photo Emission Electron Microscope) to map the plasmon nearfields on the surface of an array of the structures will be presented. It will be shown that the location and intensity of the focused nearfields can be controlled by changing the polarisation of the excitation light, enabling the switching of the plasmon energy localisation [2]. In section 4, specific symmetry and geometric properties of nanostructures will be shown to have an impact on the propagation of SPPs. In particular, it will be demonstrated and justified how in certain orientations, arrays of rotor shaped nanostructures have interesting wave-guiding interactions with propagating SPPs [3]. One result of this is a shift from P polarised illumination at which the classical farfield SPP related minimum reflectivity occurs. Following this, the first instance of plasmon mediated polarisation reorientation observed in the farfield, with no associated directional change of the farfield light, will be described and accompanied by supporting simulations [4, 5]. Finally, section 5 will deal with aspects of ultrafast dynamics of propagating SPPs. In particular, a tailor designed architecture will be examined for the possibility of generating broadband, ultrashort plasmon pulses [6]. Furthermore, the temporal modification of the illumination pulse resulting from SPP excitation will be investigated.

2. Plasmon active nanostructures and their fabrication

The field of plasmonics has taken a big leap in recent years, with one of the major attributors being the development of techniques to fabricate the micro- and nano-structures needed to control the flow and storage of electromagnetic energy on a very small scale. This local excitation and control of SPPs requires structuring techniques with nanoscale precision, of which electron-beam lithography and focused ion beam irradiation have proven to be the most important because of their ability to make diverse structures with high resolution. In this section, these structuring techniques will be briefly discussed, along with the implications of sample quality.

2.1. Nanostructuring

Electron-Beam Lithography (EBL) is a process that uses a focused beam of electrons to form patterns for material deposition on (or removal from) the sample substrate. In comparison to optical lithography, which uses light for the same purpose, EBL offers higher patterning resolution because of the shorter wavelength possessed by the 10-50 keV electrons that it employs. This small diameter focused beam of electrons is scanned over a surface, negating the need for masks required in optical lithography for the projection of patterns. An EBL system simply draws the pattern over the resist wafer using the electron beam as its drawing pen. Thus, EBL systems produce the resist pattern in a serial manner. This makes it slow compared to optical systems, but gives a user more control of the structure shape and allows for different rate of lithography at different locations.

The *Focused Ion Beam* (FIB) technique was developed during the late 1970s and the early 1980s, with the first commercial devices available in the late 1980's [7]. The technology enables localised milling and deposition of conductors and insulators with high precision, hence its success in device modification, mask repair, process control and nanopatterning [8–10]. When energetic ions hit the surface of a solid sample, they lose energy to the electrons of the solid as well as to its atoms. The most important physical effects of incident ions on the substrate are: sputtering of neutral and ionized substrate atoms (this effect enables substrate milling), electron emission (this effect enables imaging), displacement of atoms in the solid (induced damage) and emission of phonons (heating). Chemical interactions include the breaking of chemical bonds, thereby dissociating molecules (this effect is exploited during deposition). The best resolution of FIB imaging and milling is comparable to the minimum ion beam spot size, typically below 10 nm. In crystalline materials, such as aluminium and silver, the ion penetration depth varies due to channeling along open columns in the lattice structure. The removal of sample material is achieved using a high ion current beam, resulting in a physical sputtering of sample material. By scanning the beam over the substrate, an arbitrary shape can be etched.

Thin Metal Films: A basic requirement for experimental research on SPPs is the ability to make high purity, smooth, and often thin, metallic films. A number of techniques are available for this, the most typical being resistive thermal evaporation and e-beam evaporation.

The principle of vacuum evaporation is simple: the substrate and the coating material are both placed in an evacuated enclosure, some distance apart. The coating material is then heated to its vaporisation pressure point, so that it evaporates. Sufficient thermal energy is supplied to enable individual atoms to escape from the surface of the molten material. These atoms travel in a straight line through the vacuum towards the substrate where they adhere to the surface. Several methods may be used to melt the evaporant, such as resistive heating in which the evaporant is loaded in a boat shaped crucible made of a metal with a considerably higher melting point than the evaporant, through which a large DC current is passed. Many metals may be evaporated very successfully using resistive heating, and of particular importance here, this list includes Silver, Gold, Aluminium, Nickel, Platinum and Chromium; some of which are important as plasmon active substrates, and others important for adhesion layers [11].

An alternative deposition technique is electron beam heating. Here, a hot wire filament is used as a thermionic electron emitter, where the electrons are accelerated and guided toward the evaporant (either directly, or contained in a crucible). The evaporant is heated by the kinetic energy of the electrons and subsequently is coated on the sample. This method can typically generate smoother films in comparison to the resistive heating method, as the material is evaporated from its surface in a much more controlled manner than resistive heating, where the sample is typically completely melted and prone to sputtering.

In either system, the rapid condensation typically produces grained films which are relatively rough on an atomic scale, but on the scale of the wavelength of the radiation they are typically very smooth. Therefore, roughness induced re-radiation of an excited plasmon will be at an acceptably low level for the majority of plasmon experiments.

2.2. Three fold symmetric structures

Experimental examinations presented in this chapter primarily deal with arrays of structures of 120° symmetry properties, along with a reference sample of a more typical geometry (ring or doughnut shape). The geometry of four example structures are depicted in Fig. 1. They are designed such that they relinquish the widely investigated circular symmetry in favor of a 120°, or 3 fold, symmetry. One of the primary and original concepts behind this 3 fold symmetry design is that it permits the reduction of the footprint of the structure in comparison to a ring design, while maintaining surface plasmon resonance conditions. For a ring shape structure, an optimum nearfield resonance will occur where the ring diameter is an (low) integer multiple of the plasmon wavelength. If this is the case, a plasmon propagating around the structure will not destructively interfere with itself, but instead each circulation plasmon wave will constructively add to the other propagating waves. The same is true for the other structures displayed in Fig. 1, but as a result of the more complex shape, the plasmon propagation will be more complex and exhibit a smaller footprint. Additionally, new nearfield focuses will be introduced, for example at the structure centers. The structure geometries were analytically designed according to Lissajous type functions, or more specifically Epitrochoide geometries [12]. Using these functions, the geometries of the structures are defined according to the following equations:

Figure 1. Sample designs of structures (top) and extended structure design for production (bottom) [12].

$$x = (R+r)cos\phi - (r+\rho)cos\left(\frac{R+r}{r}\phi\right) \tag{1}$$

$$y = (R+r)sin\phi - (r+\rho)sin\left(\frac{R+r}{r}\phi\right) \tag{2}$$

Where ϕ is an angle between 0 and 2π, r is the radius of a circle that is rotated about the boundary of a larger circle of radius R, and ρ is the distance from the outer boundary of the smaller circle to its center. Numerous variations of the above described structures were manufactured in arrays using EBL [13], and were prepared as follows:

Following a 200°C, 1 hour bake, ZEP 520 photo-resist was spun on to silicon plates to a thickness of 100 nm. The desired structures designs were then written onto the photo-resist using a Joel JBX-6000FS/E EBL system and subsequently etched (with SF_6 and C_4FH) to reveal the desired surface profile. Following this, a 5 nm thick adhesion layer of Platinum was deposited, and finally an 80 nm thick silver film deposition, to facilitate SPP excitation, was made. The overall array size was typically 200 μm^2 to 400 μm^2 (varying with sample scale and grating constant), meaning the overall arrays were typically visible by eye; their visibility aided by their natural diffraction. In the majority of the samples, these geometries represent the shape of the raised topography. However, in one manufacturing phase, the geometries represent lowered (trough) locations resulting in "rotor" shaped nanostructures; the importance of which will be discussed in section 4 of this chapter.

2.3. Farfield optical analysis

The characterisation of the structured arrays begins with a farfield diffraction analysis, using a semi-spherical scanner [Fig. 2]. Here, the sample was mounted on a sample goniometer, and was illuminated with a polarised laser (HeNe or Ti:Sa). Prior to incidence on the sample, the laser beam was also passed through an iris to reduce the beam diameter to approximately 0.5 mm (slightly larger than the typical size of the structure arrays). Additionally, for CW laser operation, a Fresnel Rhombus was placed in the laserline between the polariser and the iris, allowing for variation of the polarisation with no intensity dependence related to a fundamental polarisation preference of the laser. The detector used (photomultiplier tube) was positioned behind an aperture which had an azimuthal angular acceptance of 0.15° [Fig. 2].

Performing farfield diffraction characterisation allows for the quantification of the quality of the sample from both a plasmon perspective, and sample quality. An example of a complete θ, ϕ farfield scan is presented in Fig. 3a. The sample examined is an array of the 3-fold symmetric shamrock shaped design (grating constant of 1.5 μm.). Here, the sample was examined under P polarisation illumination, and with the sample goniometer at angles of $\theta = 45°$, $\phi = 0°$ and $\alpha = 45°$ (diffraction pattern rotated by 45° to its normal). Series of scans such as this allow for the identification of SPP resonances (see section 4), checking the array pitch by measuring the diffraction angles, and making some sample quality checks, related to the sharpness of the diffraction and reflection channels. Using the scanner, high resolution mapping of individual diffraction orders is possible, enabling detailed characterisation of the spatial intensity distributions of the diffraction orders. Similar analysis (called spot analysis) is widely used in low energy electron diffraction (LEED) and related surface science techniques, where it permits the characterisation of the reciprocal space structure of a surface at atomic resolutions. However, the use of spot analysis is rarely used in visible laser spectroscopy techniques, as employed here. One such spatial intensity scan is presented in Fig. 3b, for a shamrock sample under P polarisation illumination, and with the sample goniometer at angles of $\theta = 55°$, $\phi = 0°$ and $\alpha = 0°$. Immediately visible from the farfield intensity map is that the spot does not have a uniform (Gaussian) spatial distribution. This spatial distribution is not generated by the laser beam profile, which is close to (as a result of the small aperture) a Gaussian profile which is spatially cut approximately half way down its wings. This was confirmed by examining the laser profile of the reflection and diffraction from a symmetric ring shaped structure. This comparative check of the farfield spatial intensity profile of the ring structure and the shamrock structure confirms that the uneven spatial profile in Fig. 3b is as a result of the nanostructures symmetry properties. This is further confirmed by performing a Fast Fourier Transform (FFT) on the image, which reveals 6 preferential symmetry directions, directly related to the 3 fold symmetry of the nanostructures [Fig. 3c].

3. PEEM as a tool for imaging plasmonic fields

In the this section, a nearfield examination of the plasmon enhanced electromagnetic fields on the above discussed reduced symmetry structures will be presented. For this, a PEEM (photo emission electron microscope) is used to map the plasmon nearfields on the surface

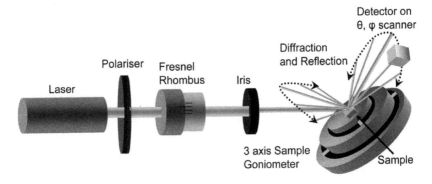

Figure 2. Set-up for farfield diffraction pattern (and individual diffraction spot) analysis [14].

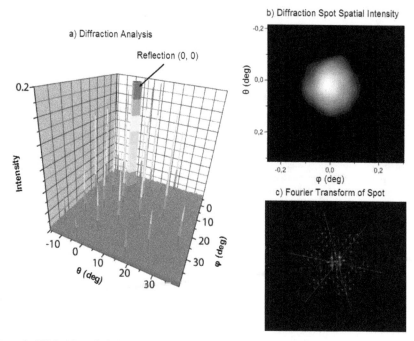

Figure 3. FIX C a) Sample diffraction characterisation. Reflection taken as the $(\theta, \phi) = (0, 0)$ point, θ and ϕ are the scan angles. b) High resolution farfield angular intensity map of single diffraction order (0, -2) from a 3-fold symmetric shamrock structure. c) Fast Fourier Transformation (FFT) of (b) demonstrates 3 preferential symmetry directions (indicated by dotted red lines) of the intensity spot, as a result of three fold symmetry of the structure [14].

of an array of threefold symmetric structures. In addition to the experimental observation of plasmon energy localization control [2], the powerful use of a PEEM for the non-perturbative subdiffraction limited imaging of plasmonic fields will be discussed.

3.1. Nearfield imaging

At the heart of much plasmon research are tools that allow researchers to examine plasmon effects in the nearfield. This was initially made possible with the development of scanning probe techniques, and their modification for the purpose of examining nearfield SPP properties directly at the surface at which the SPP is confined. The first scanning probe technique applied to the investigation of SPPs was scanning tunneling microscopy (STM), relying on the detection of changes to the tunnel current by SP induced variations in the local density of states [15–18] or the farfield scattered light due to the local SPP interaction with a STM tip [19].

Now, a wide range of techniques for the imaging of sub diffraction scale plasmon processes are available, with Scanning Nearfield Optical Microscopy (SNOM, or sometimes NSOM) being the most popular. This is a technique for optical investigation below the farfield resolution limit (diffraction limit), which is achieved by exploiting the properties of evanescent waves. The technique involves placing the detector (typically a very small probe) very close (distance smaller than the wavelength) to the specimen surface, allowing for surface inspection with high spatial resolving power. With this technique, the resolution of the image is not limited by the wavelength of the illuminating light, but rather by the size of the detector probe, along with other considerations.

Irish scientist E. H. Synge is given credit for conceiving and developing the idea for an imaging instrument that would image by exciting and collecting diffraction in the nearfield. His original idea, proposed in 1928 [20, 21], was based on the usage of intense planar light from an arc under pressure behind a thin, opaque metal film with a small aperture of about 100 nm. The aperture was to remain within 100 nm of the surface, and information was to be collected by point by point scanning. He foresaw the illumination and the detector movement being the biggest technical difficulties [22].

Current generation SNOM techniques are very powerful tools for studying SPPs; however the perturbation and alteration of the SPP field associated with the introduction of a tip is one of the major concerns and drawbacks of SNOM techniques. The use of probes (coated or uncoated fibers, or SPM tips etc) in the nearfield proximity of metal surface results in a perturbation of the electromagnetic field due to the tip / surface interaction. However, until recently, there was no known way to investigate the nearfield plasmon information without influencing the plasmon itself to some extent with the measurement device.

However, in 2005, two independent groups demonstrated the use of a PEEM for the observation of plasmonic nearfields [23, 24]. In PEEM [Fig. 4c], photo-electrons emitted from the surface of a metal are imaged with electron optics, and these electrons are collected at a distance on the millimeter scale, meaning there is no influence of the collection optics on the plasmon before or during measurements.

PEEM is closely related to the more recently developed Low Energy Electron Microscopy (LEEM), with the predominant different being that the principle of PEEM is the photoelectric effect. PEEM has already proven to be a powerful tool in material science, surface physics and chemistry, thin film magnetism, polymer science, and biology [25]. Historically, the invention of PEEM dates to the early 1930's, shortly after the introduction of electron lenses. The first working PEEM was built by Bruche in 1932, and the principal design of his PEEM is still used. In Bruche s PEEM, UV light from a mercury lamp was focused onto a sample, and the emitted photoelectrons were accelerated by a potential difference of 10 to 30 kV between the sample (cathode) and the anode of the PEEM, and subsequently focused onto a phosphor screen. A PEEM forms an image of a surface based on the spatial distribution of photoelectrons emitted. Importantly for plasmonics, photoelectron emission has been shown to be enhanced by the increase of the local electrical field upon excitation of SPPs [26].

The PEEM system we have in our lab is a SPECS PEEM P90, and depending on the photo-emission flux, this system is capable of imaging with a lateral resolution of 5 nm. The illumination frequency at which one can operate is dependent on the sample work function, which the illumination energy must be above for direct photoelectron mapping. Therefore, PEEM imaging of direct (or single) photo-emission processes presents a high resolution map of the surface workfunction threshold. For most plasmonic experiments, the surface under investigation is largely a smooth single material surface, with specifically designed sharp nanoscale structures for controlling the plasmons. On such a surface, the dominant variations in workfunction arise from the sharp topographic variations, and so these can be mapped with high resolution using single photoelectron PEEM imaging. In order to map the plasmon nearfields, multiphoton photoelectron emission is used (typically 2 Photon Photo-Emission - 2PPE). This imaging technique is particularly useful for the imaging of plasmon effects, as the local multiphoton photo-emission is extremely sensitive to local field intensities (it varies with the square of the field strength), which are dramatically stronger at plasmon localised points [23, 24, 27].

A comparison of imaging plasmons using the PEEM technique to SNOM techniques reveals a number of advantages and disadvantages:

Advantages:

First and foremost, PEEM has the ability to image the plasmon without perturbing the plasmon field. By comparison, SNOM techniques require placing a probe within the plasmon field, and so inherently altering the field.

Unlike SNOM, PEEM is not a scanning imaging technique. Like an optical microscope, it captures all field information simultaneously, but with nanometer resolution. This means that the technique is extremely quick, allowing for real time, very fast monitoring of plasmons. In fact, the imaging rate is solely dependent on the required camera integration time, which is in turn dependent on the photoelectron emission rate, and ultimately on the illumination power.

As the 2PPE process is dependent on the square of the power of the local field strength, the contrast of the PEEM image is also a squared contrast. Therefore, to a first approximation, in a 2PPE PEEM image, an area that is twice as bright as another area, actually indicates that there is only a field strength difference of the square-root of the difference between the intensities. This squared dependence on the 2PPE results in a high contrast level when compared to SNOM techniques.

Disadvantages:

The main problem with PEEM is that the work function of typical plasmon active materials restricts 2PPE to the blue end of the visible spectrum, meaning that blue excited plasmons are most appropriate. However, with higher power systems (e.g. amplified Ti:Sa systems) observation of plasmons excited at 800 nm would be possible with 3 photon photo-emission. In comparison, SNOM keeps a fundamental advantage here, where its performance increases as the excitation frequency is reduced, and significantly, it works very well at telecom frequencies.

Another disadvantage of PEEM is the associated cost of the equipment. Imaging photo-electrons requires extremely good vacuum conditions, which has a high associated cost, and also a relatively large space requirement. Additionally, the imaging optics, similar to those in a Transmission Electron Microscope (TEM) or Scanning Electron Microscope (SEM), are costly. Furthermore, for successful plasmon imaging with a PEEM, femtosecond laser systems are required for excitation; again a significant cost. A typical SNOM would cost considerably less, requiring a relatively low quality SPM as its basis (atomic scale resolution is not necessary or usable), simple CW laser light sources as illumination, and no vacuum requirements.

Despite these disadvantages of a PEEM system, the advantage and power of direct real time imaging without perturbing the plasmon field cannot be ignored.

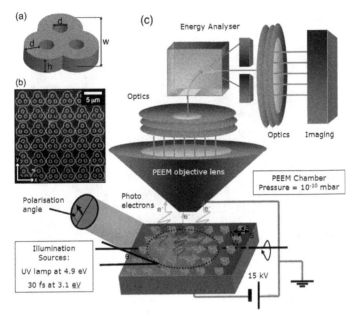

Figure 4. (a) Geometry of the structures. Structures in scale *A* have dimensions: w = 2700 nm, d = 600 nm, h = 100 nm; structures in scale *B* have dimensions w = 3600 nm, d = 800 nm, h = 100 nm. (b) SEM image of array of structures in scale *A* [2]. (c) Schematic of the technique for laser excited optical nearfield imaging with a PEEM [14]. For the presented results, the illumination angle (θ) is fixed at 25°. Note: Not all PEEMs are equipped with an energy analyser as depicted here for a Specs 90 PEEM.

3.2. Nearfield analysis of 3-fold symmetric structures

SPPs are intrinsically accompanied by strong electromagnetic nearfields, and surfaces can be actively modified to influence the excitation conditions of SPPs and hence nearfields. In particular, the selective addressability of nearfields on a surface is of interest, as shown in a demonstration of adaptive nearfield shaping [27]. In the following section, an investigation of the SPP electromagnetic nearfields excited on some of the structures described above is presented; the geometry and dimensions of which are depicted in Fig. 4. To investigate the nearfield distributions of an array of these structures, a Focus IS PEEM was used, described in detail in [28]. To record the PEEM images, two different light sources were used: a mercury-discharge lamp (UV illumination) with high-energy cutoff at 4.9 eV and a frequency-doubled Ti:Sa laser system, delivering 400 nm (3.1 eV) pulses of 100 fs duration and 20 mW at 80 MHz repetition rate. While the energy of the UV illumination is sufficient for the electrons to overcome the work function of the structured silver surface of 4.64 to 4.74 eV [29], the energy of the laser photons is too low. Hence at least two photon processes are required to generate photoelectrons; i.e. two photon photo-emission (2PPE). The laser power bandwidth product coupled to the plasmon nearfield enhancement resulted in readily observable 2PPE processes. As discussed above, as these 2PPE processes are very sensitive to the intensity of nearfields, nonlinear PEEM becomes a highly suitable tool to investigate and map plasmonic nearfield processes.

Plasmonics: Advanced Topics and Applications

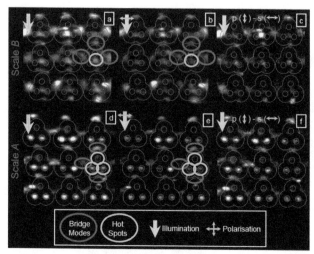

Figure 5. Contrast enhanced 2PPE PEEM images of a 3 by 3 cut out of the array of structures (a, b, d, and e). Images (c) and (f) show the differential image of S polarisation subtracted from P polarisation. 400 nm, ∼ 100 fs laser illumination incident as indicated by yellow arrows (at an angle of $\theta = 25°$. Polarisation is indicated by green arrows, P and S corresponding to vertical and horizontal respectively. For better recognition, the contour of the structures is highlighted (red). Deviations from the regular array pattern are due to spherical aberrations of the PEEM electron optics. Green circles represent hot spots as discussed in text and blue circles indicate bridge modes [2].

3.3. Plasmon energy localisation control observed by PEEM

For our presented investigation, firstly, the structures have been imaged using CW excitation at 4.9 eV to map the work function of the structure surface. These maps can be used as a reference for the 2PPE PEEM images to distinguish photo-emission effects due to the local electronic and morphologic structure of the sample surface from optical nearfields. Additionally, these work function images are used as a basis for aligning the actual structure geometries to the 2PPE PEEM images, whose size is independently confirmed by SEM (e.g. Fig. 4b) and AFM. Following this sample workfunction mapping, the structures have been investigated using the pulsed laser excitation for different illumination polarisation conditions. These 2PPE PEEM images were background subtracted and normalised via a sample independent reference beam on the microchannel plate in the imaging system of the PEEM, and subsequently contrast enhanced. In these images it is important to remember that intensity differences do not correlate linearly to the nearfield intensities, as discussed above. From these 2PPE PEEM images [Fig. 5], we can clearly identify nearfield plasmon effects. These excitations can be assigned to two different locations, at which they occur periodically across the entire array of structures:

Firstly, areas between the individual structures are excited, indicated by the bright intensities, which connect the structures of the array with each other vertically and horizontally. This emission occurs identically for structures of scales *A* and *B*, as highlighted by blue indicators in Fig. 5. These gap modes are due to field enhancement effects, which occur between the exposed edges of one structure towards its neighbors in the array.

Secondly, the other location where noteworthy enhancement takes place lies within the contour of the individual structures. These hot spots appear in each structure of the array. For scale *A* structures, these excitations are up to 280% brighter than the bridge excitations between the structures, depending predominantly on the intensity of the individual spot itself. This indicates an increased localisation of nearfield intensity in hot spots on the surface. Comparing Figs. 5a and 5d shows, additionally, that the strength and location of these excitations is dependent on structure size: for the larger structures the hot spots are located within the holes of the structure of the circles pointing along the *x* axis (there are no comparable excitations in the holes of the circles oriented in the *y* direction). Whereas, for the smaller structures the excitations are centered between the holes of the two circles along the *x* axis. Their intensity is much less pronounced and similar to the bridge excitations on the same array. Their excitation is promoted by geometric conditions resulting in the excitation of localized SPPs. The change of the location and intensity of the excitations within the contours of the individual structures is dependent on structure size and can hence be attributed to different interference and resonance conditions for the SPPs, determined purely by the geometric considerations.

To examine any polarisation dependence of these effects, P and S incident polarisations were examined for structures of both sizes. Images in Figs. 5a and 5d are acquired with P polarised light, images in Figs. 5b and 5e with S polarised illumination light. For comparison, S polarisation images are subtracted from P polarisation images [Figs. 5c and 5f]. The subtraction images demonstrate directly that the excitations within the contours of the individual structures generated with P polarisation have a stronger intensity than the ones generated with S polarisation for structures of both structure sizes. Surprisingly, the locations of the excitations are independent of the polarisation used. A quantitative analysis of the hot spots in the larger structures shows that the P intensity is about 50% higher than the S intensity. For the smaller structures the enhancement factor for the excitations within the contours of the structures varies in the range from 20% to 80%. The deviations in nearfield intensities from one structure to another can be attributed to the roughness peaks [30], but the location of the hot spots is predominantly determined by the geometric arrangement and shape of the structures.

3.4. Conclusion

A focusing of the SPP nearfields by threefold symmetrical to well defined locations is demonstrated. Observation of this effect is achieved using PEEM as a tool to map electromagnetic nearfields. The location of the focused nearfields varies with, and can be chosen by, the original structure size and design. More importantly, the intensity of locally fixed nearfields (hot spots) can be influenced by changing the polarisation of the excitation light by rotating the polarisation from S to P orientation. This enables one to switch the energy localisation on and off. Hence this approach presents a step towards a predictable design of structured surfaces, which focus energy in a spatially selective and switchable manner [2].

4. SPPs on 3-fold symmetry nanostructures

Of primary interest in this section are two arrays featuring nanostructures based on two different symmetry classes. The first array (the ring array) is designed with nanostructured rings invariant under C_∞ transformations and will be presented here as a reference array. In the second array (the rotor array), the structures resemble the shape of a triquetra rotor [Figs.

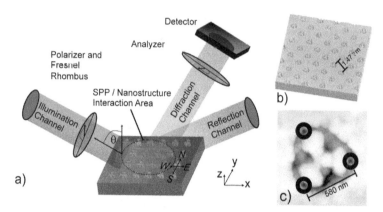

Figure 6. a) Visualisation of SPP excitation, propagation and re-emission processes, on a nanostructure arrayed surface [4]. Compass notation indicates the 4 examined interaction orientations - the displayed orientation in the schematic is E. b) AFM image of nanostructures array. c) Individual rotor structure including indication of scattering points used in the simulations.

6b and 6c]. They have a threefold symmetry and are invariant under C_3 transformations. The nanostructures are arranged to form a squared array, as indicated in Figs. 6a and 6b. As the rotor array is examined in four 90° separated orientations (α), compass notation (N, S, E, and W) will be used as identification labels [Fig. 6a]. Both arrays are housed on a 1 mm² silver coated section of silicon wafer and were prepared using e-beam lithography followed by etching to reveal the desired surface profile, as described above.

The experimental setup [Figs. 2 and 6a] for the farfield polarisation examination is as follows: A laser source ($\lambda = 632.8$ nm) is collimated, polarised (extinction ratio of 10,000:1), and made incident on the sample which is housed on a rotation table on a fine adjust goniometer [Fig. 2]. A Fresnel Rhombus is positioned in the beam-line between the polariser and the sample. This allows for polarisation angle (β) variation with uniform beam intensity, independent of any fundamental polarisation of the laser. The detector (photodiode) is mounted on a computer controlled, highly resolving, angular scanner which has the sample goniometer at its fulcrum. A polarising analyser (extinction ratio of 10 000:1, analyser angle = γ) can be positioned on the semi-spherical scanner in front of the detector, depending on the experimental requirements.

4.1. Plasmon excitation on nanostructured arrays

Certain anomalies in the intensity of light diffracted from a grating are known to correspond to the excitation of SPPs [30]. They are apparent from sharp changes in the reflected intensity of P polarised light when the grating vector is parallel to the plane of incidence. In this case, the grating changes the in-plane wave vector of the incident photon field by the addition or subtraction of integer multiples of the grating wave vector [31, 32].

In order to locate the grating induced SPPs, angular scans for S and P polarisation were recorded in a direction about an axis parallel to the S oscillation, while maintaining the sample perpendicular to the laser in the plane of P oscillation. Such angular scans were carried out on a number of available diffraction orders for both samples in all four structure orientations. Here, for consistency and clarity, we predominantly limit our presentation to the

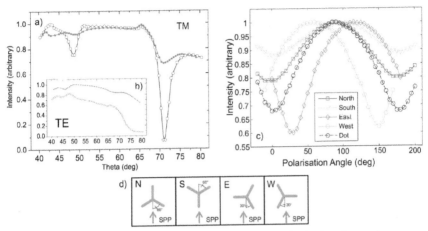

Figure 7. Angular scans of the intensity of the (+2, 0) diffraction order for the reference ring array (unfilled) and the rotor array in the North orientation (red, filled) for P (TM) polarised (a) and S (TE) polarised (b) light. c) Fixed angle polarisation scan for the ring array and for N,S,E and W orientations of the rotor array (P = 0°, 180°, S = 90°). Note: The analyser indicated in Fig. 6a is absent for the results displayed in here. d) Four orientations of the rotors primary axis with respect to the excited propagating SPP [3].

(2, 0) diffraction order (however, the effect is not limited to this particular diffraction order [14]). For the reference (ring) array, the P polarisation curve reveals two pronounced minima at 71.3° and 48.5°, where the SPP extracts energy [Fig. 7a]. From this figure, it is clear that for the ring array the coupling efficiency for P polarised light is very strong; for the SPP at 71°, $I_{SPP}/I_{max} > 10\%$. For the rotor array, the excitation efficiencies at P polarised illumination are considerably lower than for the ring array. In terms of the angular resonance scan, the only effect the nanostructure design can have is on the efficiency at which light can be coupled to the surface. Generally, for a grating with single periodicity, the further one deviates from a sinusoidal cross-section profile, the lower coupling efficiency one gets, as a Fourier analysis of the profile reveals smaller (but broader) peaks [33].

4.2. SPP illumination polarisation dependence

As the ring array is completely invariant under 0°, 90°, 180°, and 270° rotations (α), angular scans at these orientations result in an identical plot to the corresponding plot in Fig. 7a. However, as a result of the 120° symmetry of the rotor structures, no overall 90° or 180° rotational symmetry is conserved. Therefore, if the structures themselves are to have an effect on the SPP resonance conditions, examinations with the structures in N, S, E, and W orientations should present individual differences. To this extent, it is observed that the rotor structures have a definite impact on the polarisation angle at which SPP related minimum reflectivities occur [Fig. 7c]. In order to investigate this phenomenon, the arrays were mounted at an angle such that the examined diffraction order was in resonance for P polarised light. The polarisation direction was subsequently rotated using the Fresnel rhombus in 5° increments through a full polarisation rotation. In contrast to such a scan using the ring array, which shows minimum reflectivity for purely P polarised light, for the rotor array it

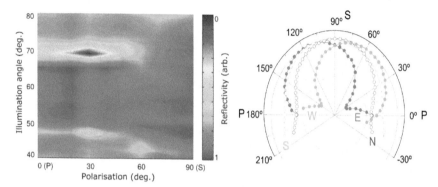

Figure 8. Left: Measured reflectivity for an illumination angular range, $\theta = 40°$ to $80°$, and illumination polarisation of P to S. Peaks at $\theta = 70°$ and $47°$ indicate plasmon modes. Right: Intensity of the ($+2^{nd}$) diffraction order as a function of polarisation angle (β) for rotor nanostructures in the 4 orientations (N, S, E, W) at the SPP excitation illumination angle ($\theta = 70°$) [4]. Intensities are individually normalised to 1. Note: The analyser indicated in Fig. 6a is absent for the results displayed in these plots.

is observed that the minimum reflectivity does not necessarily occur for incoming P polarised light [Fig. 7c]. While the minimum reflectivities for N and S orientations are found at P polarisation, this is not the case for E and W orientations, where the minimum is shifted from P polarisation by $+30°$ and $-30°$, respectively [3].

Before this illumination polarisation shift of the reflectivity minima can be confirmed to be solely a result of an interaction between propagating SPPs with the rotor structures, it must be confirmed that it is not purely a grating artifact; as it is well known that a complex grating topography can present changes in farfield intensity, independent of plasmonic effects [33, 35]. Therefore, we carried out a complete angular and illumination polarisation characterisation of our gratings for all 4 illumination orientations. An example of such an angular/polarisation scan for the SPP/rotor interaction in the E orientation is presented in Fig. 8; demonstrating that:

a) The SPP excitations at illumination angles (θ) of $70°$ and $47°$ are the only pronounced intensity variations.

b) For this SPP/nanostructure orientation (E) both plasmon reflectivity minima are shifted to a polarisation angle (β) of TM + $30°$.

From a qualitative point of view, an explanation of this shift of illumination polarisation corresponding to SPP related minimum reflectivity can be found in considering the symmetry of the structures with respect to the incoming illumination, and hence initial SPP propagation direction [Fig. 7d]. Turning our attention to processes occurring after excitation; consider the rotors in the E orientation. A grating induced SPP would propagate along the silver surface, where upon reaching the boundary of a rotor structure, the SPP wave would undergo a number of different processes with different probabilities [36]. As the SPP wave impinges on the boundary, the portion of the wave that is not transmitted or reradiated interacts with the boundary in two manners:

The first is through reflection [37, 38] and the second, and more interesting here, is through a guiding effect which can occur when a propagating SPP is made incident on a guiding surface feature [39–42]. It is anticipated that our triangular trough boundary acts much like a waveguide; behaving as a gutter collecting the SPPs, and guiding them into propagation within the trough. This behaviour can only occur with relatively high efficiencies if there is an acute angle between the original SPP direction and the new guided direction, e.g. in analogy to skimming a stone on water. Following this, the SPP can be reradiated where its polarisation would be determined by the polarisation of the originally guided wave [43, 44]. As a result of the phase shift between the two optical channels contribution to this diffracted mode [45], these two light components will predominantly destructively interfere. Naturally, this destructive interference will be at a maximum where the polarisation states of the two interfering components are matching. It is important to remember that changing the incoming illumination polarisation angle does not change the associated polarisation orientation of the excited SPP; it only alters its relative excitation strength. For the rotor structure in the E orientation, for this polarisation matching to occur, the polarisation of the incoming light would be set to P +30°; matching the twist in the SPPs associated polarisation as described above. By a similar argument, with the rotors in the W orientation, this polarisation matching would occur at P - 30°.

4.3. Plasmon mediated polarisation twisting

In order to investigate the origin and processes involved in this polarisation minimum shift, we have used a polariser/analyser set-up as in Fig. 6a. With the illumination angle set at the SPP excitation angle ($\theta = 70°$), and the illumination polarisation set to P, we have recorded the intensity monitored by the detector, as a function of analyser polarisation. Such an examination would typically present a $\cos^2$ function of the angle between the polariser and the analyser. With the exception of an intensity offset, this is exactly what is observed for illumination angles off SPP resonance, and also for the symmetric SPP/nanostructure interaction orientations (N and S). However, for such a scan in the E and W orientations at the SPP excitation angle, we observe a deviation from a $\cos^2$ function [Fig. 9 left panel]. Most notably, we observe a 5° shift in analyser angle at which we observe a maximum. This deviation from a $\cos^2$ function indicates that our plasmon-rotor interaction is causing an additional polarising function. More specifically, this is proof that this interaction is twisting the polarisation of the light involved in the SPP excitation and re-emission process.

Although in this setup we observe a shift of only 5°, instead of the 30° shift observed in Fig. 8, both results are in fact consistent with each other. The apparent difference originates from that fact that one measurement takes into account SPP and regular diffraction channel processes, while the other isolates the SPP excitation processes [4].

This plasmon mediated polarisation twisting process is further confirmed by actively altering the distribution of light following the two paths. The variation in this ratio can be accurately controlled by tuning the illumination angle (θ), as presented in the right panel of Fig. 9. This plot of observed maxima in the polariser analyser experiment as a function of illumination angle, not only confirms the process of polarisation twisting, but also demonstrates the external active control over it that can be readily achieved (steps of 0.25° twisting readily realisable). Furthermore, the range of polarisation twisting could be greatly increased by improving both structure design and surface quality, currently limiting the SPP excitation efficiency to below 20%. If this excitation efficiency is increased, for example to 50%, the farfield measurable polarisation reorientation effect would be increased to 15°, and modifying the design of the structures to re-orientate the plasmon more efficiently would also increase the effect.

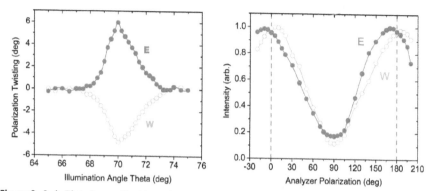

Figure 9. Left: Plot of normalised intensity as a function of analyser polarisation (γ), for illumination polarisation (β) of P. Right: Plot of polarisation twisting degree as a function of illumination angle (θ), for illumination polarisation of P [4].

4.4. Plasmon / rotor interaction simulations

To further understand the origin of the observed polarisation twisting, numerical simulations based on elastic SPP scattering [46] have been performed. In these simulations, we examine the interaction between the SPP associated electromagnetic field with the 120° symmetric structures. Initially, to allow us to focus solely on the origin of the polarisation twisting some simplifications of the processes contributing to the experimental observations have been deliberately made. This approach allows us to easily define and focus on the details we are interested in (namely the plasmon **E** field scattering) and so understand the real fundamental roots of polarisation twisting process. In these initial simulations performed by Dr. Vohnsen, a scalar effective polarisability representative of the scattering strength of each nanostructure has been used. Additionally, an ideal planar incident SPP has been assumed, absorption losses have been neglected (as these are negligible on the scale of an individual rotor structure), and for simplicity multiple SPP scattering between the structures has been omitted. By implementing these simplifications, a focus can be made on the origin of the plasmon **E** field re-orientation, which ultimately determines the polarisation re-orientation. Figs. 10a and 10d show results obtained with an individual rotor nanostructure in the N and W orientations and illuminated by an incident SPP from below. For the SPP incident in the W orientation, a change in the main **E** field direction of approximately 3° with respect to the direction of incidence is observed. This is also observed where a array of the structures is considered, as shown in Fig. 10e. This is caused by the interference of the incident SPP and the asymmetric configuration of the three-particle rotor model. In comparison to Figs. 10a and 10b where no resultant redirection of the **E** field occurs, the redirection of the plasmon visible in Figs. 10d and 10e implies that the propagation direction of the plasmon on the surface is redirected. However, the grating conditions place strict restrictions on the direction (or channels) at which light can leave the grating. Indeed, it is the additional conditions imposed by the grating that enables a polarisation twisting that, importantly, is not accompanied by a farfield relocation (spatial shift) of the light. This is typically not the case, where the polarisation *and* spatial conditions of the light are defined by the plasmon **E** field prior to reradiation. Therefore, for our structures, the polarisation of the reradiated light is defined by the plasmon **E** field immediately prior to the reradiating process, but the spatial direction of the light is defined by the grating conditions. For this reason, we can label the effect we observe as a true polarisation twisting.

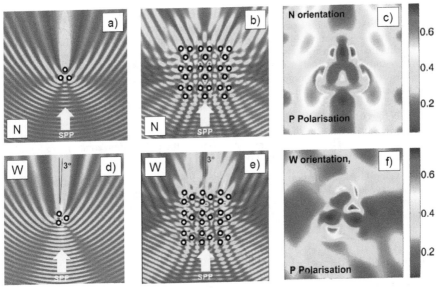

Figure 10. a&d) 10 x 10 micron field amplitude images of the incident and elastic scattered lossless SPPs, interacting with individual rotor representative structures in the N (a) and W (d) orientations [4]. b&e) Extension to interaction of SPPs with 3 by 3 arrays of the structures in the N (b) and W (e) orientations [14]. c&f) FDTD simulated nearfield energy distribution for the rotor structures in N (c) and W (f) orientations for P polarisation [5].

The difference in the degree of polarisation twisting between the experimental observations and the simulations is accounted for in the deliberate simplifications made in the simulations; especially in the substitution of the complex structures with just 3 scattering points. Regardless of these simplifications, the primary function of these simulations here is to identify and understand the dominating mechanisms contributing to the experimentally observed polarisation twisting. This has been confirmed to be as a result of an asymmetrical in-plane SPP scattering while the SPP/rotor interaction is in specific orientations (E and W). However, full FDTD calculations (by the group of Prof. Runge) demonstrate the experimentally observed 30° reorientation [5]. In these simulations, the near- and far-field properties of the rotor structure in the time-domain is calculated using the program FDTD solutions of Lumerical Solutions Inc. From these simulations, the electromagnetic nearfields, and the farfield reflectivities are calculated, as displayed in Figs. 10c and 10f. Importantly, as with the elastic SPP scattering simulations discussed above, in Fig. 10f a reorientation of the plasmon **E** field is again observed, but for these more complete simulations, the true degree of reorientation of ($\sim 30°$) is apparent, corroborating the experimental findings presented above.

4.5. Conclusion

In this section, it has been shown that symmetry properties nanostructures can be designed to control the propagation of SPPs on the surface. In particular, it has been demonstrated and justified how in certain orientations, rotor shaped nanostructures have interesting wave-guiding interactions with propagating SPPs, resulting in a shift from P polarised illumination at which the farfield SPP related minimum reflectivity occurs [3]. Building on this, the first instance of plasmon mediated polarisation reorientation observed in the farfield with no associated directional change of the farfield light was described. For this, an experimental demonstration of how tailor designed topographic structures of threefold symmetry can be used to alter the polarisation of an EM wave by a selective amount was made. It was isolated and confirmed that the primary process involved in this polarisation twist is from the interaction of a propagating SPP wave with the nanostructures. Specifically, the polarisation orientation of the light is determined by the E field orientation of the plasmon directly before its re-emission, and the farfield spatial location is determined by the grating conditions. This results in the observed polarisation twisting with no associated farfield directional change. The only apparent restrictions on the polarisation rotation are found to be the initial plasmon excitation conditions [4].

Finally, using Green's function based simulations, the interaction between a propagating plasmon wave and 120° three-fold symmetric structures was examined, confirming that the origin of the farfield polarisation twisting is an asymmetrical in-plane SPP scattering occurring in the nearfield. This computational observation is further confirmed by FDTD calculations of the same structures [5].

5. Ultrafast broadband plasmonics

Time scales associated with SPPs vary from 100's of attoseconds to 100's of nanoseconds. The lower limit is a theoretical limit defined by the inverse spectral width of a broadband plasmonic resonance [47], and is one of the fastest time scales in optics. However, as of yet, little experimental output has come from examining freely propagating plasmons at metal / dielectric interfaces on these ultrashort time scales. The reason for this is as a result of the combination of the difficulty in making accurate measurements on a suitable time scale, but more importantly, the difficulty in accessing plasmon modes of suitable bandwidth to support these ultrafast processes. Despite these difficulties, understanding these ultraquick processes is of key importance to the field of nanoplasmonics, and could have potential applications in, for example, ultrafast computations, and data control and storage on the nanoscale. Recently, a system that provides access to the efficient excitation of broadband, propagating plasmon modes, capable of supporting SPP pulses with temporal lengths on the 20 fs scale has been designed [6]. To achieve this feat, a surface array of tailor designed, reduced symmetry nanostructures has been specifically architected to enable the appropriate control of the plasmon dispersion relation.

5.1. Ultrafast plasmonics

The vast majority of the experimental work on ultrafast SPP dynamics have either dealt with temporal dephasing of particle plasmons [48–53], or have been aimed at understanding the processes through which unexpected levels of optical transmission in subwavelength perforated thin metal films occurs. Indeed, for the ultrafast dynamics of SPPs, the major focus of experimental research has been geared toward the understanding of extraordinary optical transmission (EOT); first observed in the visible regime by Ebbesen et al. [31]. The complete underlying processes of EOT is still somewhat of a debate [32, 54–57]; and as a result, in order to understand the process of EOT in more detail, researches have examined the temporal characteristics of EOT using ultrashort pulse illumination.

The first of these experimental examinations was carried out by the group of Ebbesen [58] where the transit time of a $\sim$100 fs pulse passing through a metal film perforated with an array of subwavelength apertures was considered. Light transmitted through a subwavelength aperture was coherent with the incident pulse, and showed a 7 fs total transient time over the 0.3 mm layer of silver film. The authors report that these delay times support the general picture in which the resonant coupling of light with metallic surface modes is responsible for the relatively large transmission and the slow group velocity of light inside these subwavelength apertures. In the following years, further theoretical and experimental examinations were made based on similar designs, but instead of assigning this delay to the finite transit time for light propagation through the nanoholes, it was assigned to the SPP lifetime for such a structure. In a theoretical work studying the propagation of 10 fs pulses (shorter than the damping time of SP excitations at the interfaces of the metal film), pronounced temporal oscillations in the transmitted light were predicted [59]. It was concluded that these oscillations reflect the temporal character of the coupling of SPPs at both interfaces via photon tunneling through the nanohole channel. More recent experiments have also confirmed a modification of the dephasing rates due to interactions between localised particle plasmons and optical waveguide modes, and subsequent modification of the photonic density of states [60]. In [61] it was demonstrated that SP transmission peaks through the nanohole arrays were homogeneously broadened by the SP radiative lifetime. From the same group an experimental study of ultrafast light propagation through plasmonic nanocrystals using light pulses much shorter in duration than the SPP damping time, was made [62]. Here, phase-resolved measurements of the time structure of the transmitted light allowed for the identification of two different contributions to the EOT effect to be nonresonant tunneling and SPP re-radiation.

Other recent reports of specific significance on the topic of ultrafast SPP dynamics have used pump probe experiments, combined with SP induced photoelectron imaging (PEEM) to achieve nanoscale spatial, and femtosecond (even sub fs) temporal imaging. In [63], ultrafast laser spectroscopy and PEEM were combined to image the quantum interference of localised SP waves. This technique permitted imaging of the spatio-temporal evolution of SP fields with a 50 nm spatial resolution taken at a 330 as frame rate. Using a similar technique, SP dynamics in silver nanoparticles have been studied [26], and an investigation of the optical nearfield was demonstrated by mapping photoemitted electrons from specially tailored Ag nanoparticles deposited on a Si substrate [23]. While on the topic of active plasmon control, femtosecond optical switching of a propagating SPP signal was reported in [64]. Here, experimental examination and theoretical analysis show that femtosecond plasmon pulses can be generated, transmitted, modulated and decoupled for detection in a single device.

Regarding the future of ultrafast plasmonics; a number of publications have recently been made indicating some of the potential directions of nanoscale spatial and fs temporal plasmonics research [47, 65, 66]. One of the key requirements to achieve the potential that ultrafast plasmonics can offer is the ability to access plasmon modes of suitable bandwidth; this will be the focus of the next section.

5.2. Broadband ultrashort propagating plasmon pulses

For a typical Ti:Sa laser system, following group delay dispersion compensation, near transform limited (sub 20 fs) illumination pulses can be generated. In order to permit the excitation of SPP pulses of comparable temporal duration to such illumination pulses, the first requirement is that the SPP excitation mechanism simultaneously envelopes the

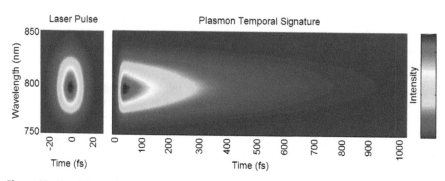

Figure 11. Simulation of the temporal signature of a broadband propagating SPP (right) driven by a sub 20 fs broadband laser pulse from a Ti:Sa laser (left).

complete spectrum of the illumination pulse. In order to spectrally envelope a sub 20 fs pulse bandwidth, an SPP coupling acceptance bandwidth exceeding 80 nm is required. Furthermore, for propagating plasmons, the illumination pulse must remain spatially and temporally optimized prior to interaction with the surface. This implies that the required broadband excitation mechanism must couple this complete illumination bandwidth at a single angle of incidence. However, for high efficiency SPP excitation architectures (i.e. typical grating or attenuated total reflection coupling) the SPP dispersion relation varies rapidly with illumination frequency and angle [30]. For example, in an ATR configuration, the full bandwidth of a spatially and temporally unchirped ultrashort pulse cannot be coupled to an SPP simultaneously [69], thus prohibiting the generation of SPP pulses of comparable duration to the driving ultrashort laser pulse. For some nanofeature based excitation mechanisms (e.g. nanoparticle, rough surface [30], slit [70], etc.) a suitable broad range of momenta can be inherently provided, allowing for broadband plasmon excitation. However, such excitation schemes are limited to comparably weak SPP generation [70], with only a very small percentage of the illumination light coupled to the desired SPP mode. Furthermore, such coupling techniques are not suitable for propagating plasmon pulse generation, as the excited plasmons will have a range of group and phase velocities, will not co-propagate, and so will be strongly spatially and temporally chirped. With these restrictions in mind, a primary research objective is to overcome these obstacles and realize a highly efficient broadband SPP excitation mechanism.

For the laser excitation wavelength of a Ti:Sa laser centered at 800 nm [67], the unperturbed SPP lifetime is $\sim$ 230 fs for a silver / air interface. A theoretical prediction of the temporal evolution of a broadband SPP excited by a 20 fs broadband pulse is presented in Fig. 11. This simulation shows how a plasmon propagating on a silver surface would evolve and decay in time. It also demonstrates that on this planar silver/dielectric interface, the SPP dispersion relation implies that as the plasmon pulse propagates, its wavefront will become distorted due to the fact that the $1/e$ decay time ranges from approximately 155 fs at the blue end of the Ti:Sa spectrum (750 nm) to 260 fs at the red end (850 nm). This introduces an additional restriction on accessing non-distorting ultrashort plasmon pulses; as it means that a broadband plasmon propagating at a planar metal interface will always become chirped as it propagates.

Therefore, one of the primary challenges in ultrafast plasmonics is the design of a plasmon excitation mechanism that allows for the efficient excitation of broadband, non-distorting plasmon modes. As justified above, this feat will not be possible for planar interface

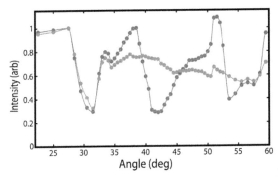

Figure 12. Intensity reflectivity scans for a 800 nm CW laser (red) and a broadband (centered at 800 nm) sub 20 fs laser (blue). Three pronounced plasmonic modes are apparent for CW light located at 31°, 42° and 55°, but only one pronounced mode is visible for the broadband pulse at 31°.

arrangements. Thus, in order to achieve this goal, our experiments deal with SPP excitation on an array of reduced symmetry nanostructures. The initial design for the structures was conceived and optimized using reciprocal space analysis. The geometry was chosen to present a range of inter-structure distances to suit the near Gaussian spectral distribution of the laser, and the physical dimensions of the surface features were optimized for SPP excitation in the near IR [6].

Confirmation of SPP excitation in these tailor designed nanostructured arrays for CW light (at 800 nm) and sub 20 fs pulses of broadband light (at 760-840 nm) is presented in Fig. 12. Three SPP modes are clearly identifiable (as reflectivity minima) for the CW laser excitation at 31°, 42° and 55°. However, for the broadband source, only one sharp SPP mode at 31° is apparent. Importantly, because the plasmon mode at 31° appears identical for both broadband and CW light sources, this indicates that this mode has at least a comparable bandwidth to the broadband source. For such grating excited plasmons, it is typically not possible to excite SPPs of comparable bandwidth to a sub 20 fs Ti:Sa laser. This is because of a strong momentum variation typically exhibited by plasmon modes over such a broad spectrum; as is the case for the modes at 42° and 55° observed only for the CW scan in Fig. 12. However, for SPP excitation on the array of reduced symmetry nanostructures it is found that a plasmon mode whose angular variation over the spectral range is very low is accessible [6]. Indeed, this mode has been found to be suitable for broadband plasmon excitation, coupling the complete illumination spectrum of a 17.5 fs, Ti:Sa system. Further angular and spectral examinations of the excitation of SPPs confirm the presence of multiple plasmonic modes for the CW source; but importantly they also reveal multiple modes for the broadband source. These additional modes are found to have the typical strong spectral variation [30], and so are not suitable for broadband SPP excitation. However, for the mode at 31° plasmons of very high bandwidth (and so very short temporal characteristics) can be excited [6].

5.3. Conclusion

In final section, a method for the excitation of broadband plasmonic modes in the near-IR regime was presented [6]. This has been achieved using a tailor designed reduced symmetry periodic surface that grants access to an SPP mode which has a fixed momentum value over the entire bandwidth of an ultrafast Ti:Sa laser. For this sample, as a result of the well ordered array basis, a high SPP coupling efficiency is achieved, and as a result of the reduced symmetry nanostructures, the range of momenta provided by the grating in a fixed direction is increased, presenting the possibility of accessing efficient, co-propagating, broadband SPPs. This ability to generate broadband, ultrashort SPP pulses that exhibit no spatial or temporal chirping in their excitation is an important step toward accessing the previously predicted ultraquick optical processes associated with SPPs.

Acknowledgements

The majority of the work presented here is based on sections of the PhD thesis of Dr Brian Ashall, carried out under the supervision of Dr Dominic Zerulla.

The authors would like to thank Dr Michael Berndt, Prof. Martin Aeschlimann, Dr Brian Vohnsen, Prof. Erich Runge, Dr Jose Francisco Lopez-Barbera and Dr Stephen Crosbie for their valuable contributions. The authors acknowledge Science Foundation Ireland and Enterprise Ireland for ongoing funding of our research in the field of plasmonics.

Author details

Brian Ashall
School of Science, Technology, Engineering and Mathematics, Institute of Technology Tralee, Ireland

Dominic Zerulla
School of Physics, College of Science, University College Dublin, Belfield, Dublin 4, Ireland

6. References

[1] H. A. Atwater, The Promise of Plasmonics, Scientific American, April Issue (2007).
[2] M. Berndt, M. Rohmer, B. Ashall, C. Schneider, M. Aeschlimann, and D. Zerulla, Opt. Lett., 34:959 (2009).
[3] B. Ashall, M. Berndt, and D. Zerulla, Appl. Phys. Lett., 91: 203109 (2007).
[4] B. Ashall, B. Vohnsen, M. Berndt, and D. Zerulla, Phys. Rev. B, 80:245413 (2009).
[5] D. Leipold, S. Schwieger, B. Ashall, D. Zerulla, and E. Runge, Photonics and Nanostructures 8 297 (2010).
[6] B. Ashall, J. F. Lopez-Barbera, and D. Zerulla, Manuscript submitted to New Journal of Physics (2012).
[7] J. Melngailis, J. Vac. Sci. Technol. B, 5, 469 (1987).
[8] D. K. Stewart, A. F. Doyle, and J.D. Jr Casey, Proc. SPIE, 276, 4337 (1995).
[9] B. W. Ward, N. P. Economou, D. C. Shaver, J. E. Ivory, M. L. Ward, and L. A. Stern, Proc. SPIE, 92, 923 (1988).
[10] T. H. Taminiau, R. J. Moerland, F. B. Segerink, L. Kuipers, and N. F. van Hulst, Nano Lett., 7, 28 (2007).
[11] X. Jiao, J. Goeckeritz, S. Blair, and M. Oldham, Plasmonics, 4, 37 (2009).
[12] M. Berndt, Anregung von Oberflächenplasmonen auf mesoskopischen Strukturen, Diploma Thesis, Heinrich-Heine-Universität, Düsseldorf, (2007).
[13] EBL system available via the National Access Program, Tyndall National Institute in Cork.
[14] B. Ashall, Surface Plasmon Polaritons on Nanostructures Surfaces, Doctor of Philosophy Thesis, University College Dublin (2009).

[15] R. Möller, U. Albrecht, J. Boneberg, B. Koslowski, P. Leiderer, and K. Dransfeld, J. Vac. Sci. Technol. B, 9, 506 (1991).

[16] N. Kroo, J.P. Thost, M. Völcker, W. Krieger, and H. Walther, Europhys. Lett., 15, 289 (1991).

[17] D.W. Pohl and D. Courjon, Near Field Optics, Kluwer, (1993).

[18] I. I. Smolyaninov, A. V. Zayats, and O. Keller, Phys. Lett. A, 200, 438 (1995).

[19] M. Specht, J.D. Pedarnig, W.M. Heckl, and T.W. Hansch, Phys. Rev. Lett., 68, 476 (1992).

[20] E. H. Synge, Phil. Mag., 6, 356 (1928).

[21] E. H. Synge, Phil. Mag., 13, 297 (1932).

[22] B. Hecht, B. Sick, U. P. Wild, V. Deckert, R. Zenobi, O. J. F. Martin, and D. W. Dieter, J. Chem. Phys., 18, 112 (2000).

[23] M. Cinchetti, A. Gloskovskii, S. A. Nepjiko, G. Schönhense, H. Rochholz, and M. Kreiter, Phys. Rev. Lett., 95, 047601 (2005).

[24] A. Kubo, K. Onda, H. Petek, Z. Sun, Y. S. Jung, and H. K. Kim, Nano Letters, 5, 1123 (2005).

[25] J. Feng and A. Scholl, Science of Microscopy - Chapter 9: Photoemission Electron Microscopy, Springer New York (2007).

[26] J. Lehmann, M. Merschdorf, W. Pfeiffer, A. Thon, S. Voll, and G. Gerber, Phys. Rev. Lett., 85, 2921 (2000).

[27] M. Aeschlimann, M. Bauer, D. Bayer, T. Brixner, F. J. García de Abajo, W. Pfeiffer, M. Rohmer, C. Spindler, and F. Steeb, Nature, 446, 301 (2007).

[28] M. Munzinger, C. Wiemann, M. Rohmer, L. Guo, M. Aeschlimann, and M. Bauer, New J. Phys., 7, 68 (2005).

[29] H. B. Michaelson, J. Appl. Phys, 48, 4729 (1977).

[30] H. Raether, Surface Plasmons on Smooth and Rough Surfaces and on Gratings, Springer Verlag, Berlin (1988).

[31] T. W. Ebbesen, H. J. Lezec, H. F. Ghaemi, T. Thio, and P. A. Wolff, Nature, 391, 667 (1998).

[32] H. F. Ghami, T. Thio, D. E. Grupp, T. W. Ebbesen, and H. J. Lezec, Phys. Rev. B, 58, 6779 (1998).

[33] R. Petit, Electromagnetic Theory of Gratings, Springer Berlin, Springer Topics in Current Physics Vol 22 (1980).

[34] S. Rehwald, M. Berndt, F. Katzenberg, S. Schwieger, E. Runge, K. Schierbaum, and D. Zerulla, Phys. Rev.B, 76, 085420 (2007).

[35] C.H. Wilcox, Scattering Theory for Diffraction Gratings, Springer Berlin, Springer Applied Mathematical Sciences Vol 46 (1984).

[36] G. I. Stegeman, A. A. Maradudin, and T. S. Rahman, Phys. Rev. B, 23, 2576 (1981).

[37] R. F. Wallis, A. A. Maradudin, and G. I. Stegeman, Appl. Phys. Lett., 42, 764 (1983).

[38] P. Dawson, F. de Fornel, and J-P. Goudonnet, Phys. Rev. Lett., 72, 2927 (1994).

[39] Y. Satuby and M. Orenstein, Appl. Phys. Lett., 90, 251104 (2007).

[40] S. I. Bozhevolnyi, J. Erland, K. Leosson, P. M. W. Skovgaard, and J. M. Hvam, Phys. Rev. Lett., 86, 3008 (2001).

[41] S. I. Bozhevolnyi, V. S. Volkov, K. Leosson, and A. Boltasseva, Appl. Phys. Lett., 79, 1076 (2001).

[42] B. Steinberger, A. Hohenau, H. Ditlbacher, A. L. Stepanov, A. Drezet, F. R. Aussenegg, A. Leitner, and J. R. Krenn, Appl. Phys. Lett., 88, 094104 (2006).

[43] G. Isfort, K. Schierbaum, and D. Zerulla, Phys. Rev. B, 73, 033408 (2006).

[44] G. Isfort, K. Schierbaum, and D. Zerulla, Phys. Rev. B, 74, 033404 (2006).

[45] J. M. Pitarke, V. M. Silkins, E. V. Chulkov, and P. M. Echenique, Rep. Prog. Phys., 70 (2007).

[46] S. I. Bozhevolnyi and V. Coello, Phys. Rev. B, 58, 10899 (1998).

[47] M. I. Stockman, M. F. Kling, U. Kleineberg, and F. Krausz, Nat. Photon., 1, 539 (2007).

[48] C. Sönnichsen, T. Franzl, T. Wilk, G. von Plessen, J. Feldmann, O. Wilson, and P. Mulvaney, Phys. Rev. Lett., 88, 077402 (2002).

[49] B. Lamprecht, J. R. Krenn, A. Leitner, and F. R. Aussenegg, Phys. Rev. Lett., 83, 4421 (1999).

[50] B. Lamprecht, A. Leitner, and F. R. Aussenegg, Appl. Phys. B, 68, 419 (1999).

[51] A. Wokaun, J. P. Gordon, and P. F. Liao, Phys. Rev. Lett., 48, 957 (1982).

[52] T. Klar, M. Perner, S. Grosse, G. von Plessen, W. Spirkl, and J. Feldman, Phys. Rev. Lett., 80, 4249 (1998).

[53] J. Bosbach, C. Hendrich, F. Stietz, T. Vartanyan, and F. Trager, Phys. Rev. Lett., 89, 257404 (2002).

[54] U. Schroter and D. Heitmann, Phys. Rev. B, 58, 15419 (1998).

[55] J. A. Porto, F. J. Garcia-Vidal, and J. B. Pendry, Phys. Rev. Lett., 83, 2845 (1999).

[56] T. Thio, H. F. Ghaemi, H. J. Lezec, P. A. Wolff, and T. W. Ebbesen, J. Opt. Soc. Am. B, 16, 1743 (1999).

[57] M. M. J. Treacy, Appl. Phys. Lett., 75, 606 (1999).

[58] A. Dogariu, T. Thio, L. J. Wang, T. W. Ebbesen, and H. J. Lezec, Opt. Lett., 26, 450 (2001).

[59] R. Müller, V. Malyarchuk, and C. Lienau, Phys. Rev. B, 68, 205415 (2003).

[60] T. Zentgraf, A. Christ, J. Kuhl, and H. Giessen, Phys. Rev. Lett., 93, 243901 (2004).

[61] D. S. Kim, S. C. Hohng, V. Malyarchuk, Y. C. Yoon, Y. H. Ahn, K. J. Yee, J.W. Park, J. Kim, Q. H. Park, and C. Lienau, Phys. Rev. Lett., 91, 143901 (2003).

[62] C. Ropers, Müller, C. Lienau, G. Stibenz, G. Steinmeyer, D-J. Park, Y-C. Yoon, and D-S. Kim, Ultrafast dynamics of light transmission through plasmonic crystals, International Conference on Ultrafast Phenomena, Niigata, Japan (2004).

[63] A. Kubo, K. Onda, H. Petek, Z. Sun, Y. S. Jung, and H. K. Kim, Nano Lett., 5, 1123 (2005).

[64] K. F. MacDonald, Z. L. Sámson, M. I. Stockman, and N. I. Zheludev, Nature Photonics, 3, 55 (2008).

[65] M. I. Stockman, N. J. Phys., 10, 025031 (2008).

[66] J. Lin, N. Weber, A. Wirth, S. H. Chew, M. Escher, M. Merkel, M. F. Kling, M. I. Stockman, F. Krausz, and U. Kleineberg, J. Phys.: Condens. Matter, 21, 314005 (2009).

[67] Griffin C Ti:Sa Laser, Kapteyn-Murnane Laboratories Inc.

[68] R. Trebino, Frequency-Resolved Optical Gating: The Measurement of Ultrashort Pulses, Kluwer Academic Publishers (2000).

[69] S. E. Yalcin, Y. Wang, and M. Achermann, Appl. Phys. Lett., 93, 101103 (2008).

[70] J. Chen, Z. Li, M. Lei, S. Yue. J. Xiao, and Q. Gong, Opt. Exp., 19, 26463 (2011).

Chapter 5

Plasmonic Lenses

Yongqi Fu, Jun Wang and Daohua Zhang

Additional information is available at the end of the chapter

1. Introduction

The resolution of almost all conventional optical system is indispensably governed by the diffraction limit. This resolution limit can be overcome by use of focusing the evanescent waves in the near field region. The concept of "superlens" was proposed firstly by Pendry in 2000 [1]. When ε= -1and μ= -1, the negative refractive index material plate can be a perfect lens [2-4]. Because of the dispersion and absorption in the materials, the conditions of ε= -1and μ= -1 is hard to satisfy for the natural materials. Although the perfect lens may not exist, the superlens which can provide higher resolution beyond the diffraction limit have been proved. And focusing by means of surface plasmon polarisons (SPPs) by plasmonic lens is attracting much interest recently due to its unique feature of extraordinary enhanced transmission [5-8]. It means that we can focus the evanescent components of an illuminated object in the near-field region with subdiffraction-limit resolution [9]. This allows them to break the conventional barrier of diffraction limit, and leads to the formation of concentrated sub-wavelength light spots on the order of nanometers. Plasmonic lens is always consisted by metal and dielectric and can excite SPPs and always can be used for focusing, imaging, and beam shaping and so on.

In this paper, a literature review is given for the purpose of displaying a physical picture of plasmonic lenses for the relevant reader. Firstly, the basic theory about the plasmonic lens is presented. Then several examples of plasmonic lens are given. Here we mainly focus on the typical concepts of the plasmonic lens reported so far.

2. Plasmonic lens on the basis of negative refractive index materials

2.1. Superlens

Although perfect lens proposed by Pendry may not exist, superlens is realized and proved by Zhang's group in 2003 [10-13] and other research groups [14-21]. Here we mainly introduce the typical works which were done by Zhang's group. They showed that optical evanescent waves could indeed be enhanced as they passed through a sliver superlens. Figure 1 below shows configuration of the superlens they designed.

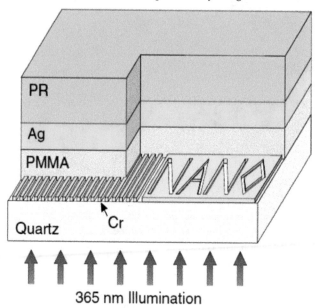

Figure 1. Optical superlensing designed by Zhang's group. Reprinted with permission from "N. Fang, H. Lee, C. Sun and X. Zhang, Science 308, 534–537 (2005)." of copyright ©2005 American Institute of Physics.

As can be seen, a set of objects were inscribed onto a chrome screen. The objects were designed to be placed about 40 nm away from the silver film which is 35 nm in thickness. And the chrome objects were patterned on quartz by using focused ion beam (FIB) technique, a 40 nm thick layer of polymethyl methacrylate (PMMA) was used to planarized them. The objects are imaged onto the photoresist on the other side of the silver film under ultraviolet (UV) illumination (at a wavelength of 365 nm). The negative photoresist which is 120 nm thick is used to record the near-field image. The substrate is food-exposed under an I-line (365 nm) mercury lamp. The exposure flux is 8 mW/cm², and an optimal exposure time of 60 s is applied to reduce the surface root mean square modulation below 1 nm for both the silver and PMMA surfaced; otherwise, the dispersion characteristics of the superlens would be modified and would in turn smear the details of the recorded image. The optical image is converted into topographic modulations by developing the negative photoresist and is mapped using atomic force microscopy (AFM).

Because the electric and magnetic responses of materials were decoupled in the near field, only the permittivity needs to be considered for transverse magnetic (TM) waves. This makes noble metals such as silver natural candidates for optical superlensing. Silver is chosen here. As surface charges accumulate at the interface between the silver and the imaging medium, the normal component of an electric of silver is selected and the permittivity of the silver and that of the adjacent medium are equal and of opposite sign. Such a delicate resance is essential to ensure the evanescent enhancement across the slab. For enhanced transmission of evanescent waves, it is found that an asymptotic impedance match ($k_{zi} / \varepsilon_i + k_{zj} / \varepsilon_j = 0$) has to be met at the surface of the silver, known as the surface plasmon excitation condition (k_{zi}, cross-plane, wave vector in silver; ε_i, permittivity of silver; k_{zj}, cross-plane wave vector in dielectric; and ε_j, permittivity of dielectric). It is widely known in metal optics that when the two media take the opposite sign in permittivity and $|\varepsilon_i| \gg \varepsilon_j$, only surface plasmons at the narrow range of in-plane wave vector (k_x) that are close to k_o can be resonantly coupled. However, less well known is that when $|\varepsilon_i| \sim \varepsilon_j$ and we are of opposite sign, the excitable surface plasmon band of k_x is significantly broadened, resulting in the superlensing effect.

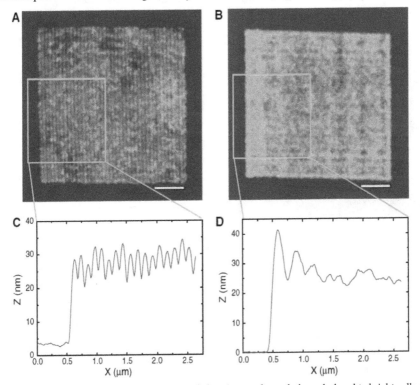

Figure 2. (A) AFM of the developed image (scale bar, 1 μm; color scale from dark red to bright yellow, 0 to 150 nm). (B) Control experiments were carried out, in which the silver superlens layer was replaced by a 35 nm thick PMMA layer, for a total PMMA thickness of 75nm. (C) The blue solid curve shows the clearly demonstrating the 63 ± 4 nm half-pitch resolved with a 35 nm silver superlens. X direction is relative displacement along the cross-section direction. (D) The blue solid curve shows the average cross section of Fig. 2B (control sample). Reprinted with permission from "N. Fang, H. Lee, C. Sun and X. Zhang, Science 308, 534-537 (2005)." of copyright ©2005 American Institute of Physics.

The intensity of evanescent waves decays with a characteristic length Z, and

$$Z^{-1} = 4\pi\sqrt{a^{-2} - \varepsilon\lambda^{-2}}\,,$$ (1)

where a is the period of a line array, and ε is the permittivity of the surrounding media. In Zhang's experiment, for the 60 nm half-pitch and $\varepsilon = 2.4$, the decay length is estimated to be 11 nm. Thus it is obviously difficult to resolve a 60 nm half-pitch object from a distance of 75 nm away if there isn't a superlens to enhance and transmit the evanescent waves. So we could find the photoresist images with typical average height modulations of 5 nm to 10 nm from Fig. 2 C. And this is assisted by careful control of the surface morphology of the PMMA and silver surface. In addition, Zhang also proved that the silver superlens can also image arbitrary nanostructures with sub-diffraction-limited resolution. The recorded image "NANO" in Fig. 3 B shows that the fine features from the mask showed in Fig. 3 A in all directions with good fidelity can be faithfully produced.

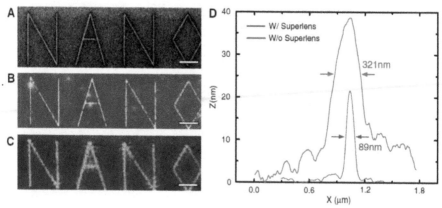

Figure 3. An arbitrary object "NANO" was imaged by silver superlens. (A) FIB image of the object. The linewidth of the "NANO" object was 40 nm. Scale bar in (A) to (C), 2 μm. (B) AFM of the developed image on photoresist with a silver superlens. (C) AFM of the developed image on photoresist when the 35 nm thick layer of silver was replaced by PMMA spacer as a control experiment. (D) The average cross section of letter "A" shows an exposed line width of 89 nm (blue line), whereas in the control experiment, we measured a diffraction-limited full width at half-maximum ling width of 321 ± 10 nm (red line). Reprinted with permission from "N. Fang, H. Lee, C. Sun and X. Zhang, Science 308, 534-537 (2005)." of copyright ©2005 American Institute of Physics.

2.2. Hyperlens

The images imaged by the superlens we talked about above are the same size as the objects. And there is no working distance. The hyperlens here was also proposed by Zhang's group [22]. It can magnify a sub-diffraction-limited image and projects it into the far field. Figure 4 A is the schematic of the hyperlens. It consists of a curved periodic stack of Ag (35 nm) and Al_2O_3 (35 nm) deposited on a half-cylindrical cavity fabricated on a quartz substrate. Sub-diffracion-limited objects were inscribed into a 50-nm-thick chrome layer located at the inner surface (air side). The anisotropic metamaterial was designed so that the radial and tangential permittivities have different signs.

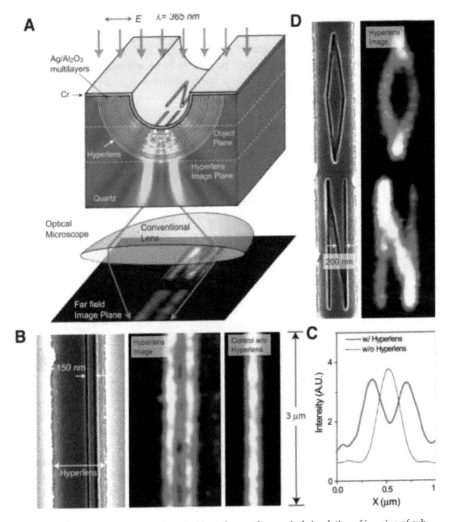

Figure 4. Optical hyperlens. (A) Schematic of heperlens and nymerical simulation of imaging of sub-diffraction-limited objects. (B) Hyperlens imaging of line pair object with line width of 35 nm and spacing of 150 nm. From left to right, scanning electron microscope image of the line pair object fabricated near the inner side of the hyperlens, magnified hyperlens image showing that the 150-nm-spaced line pair object can be clearly resolved, and the resulting diffraction-limited image from a control experiment without the hyperlens. (C) The averaged cross section of hyperlens image of the line pair object with 150-nm spacing (red), whereas a diffraction-limited image obtained in the control experiment (green). A.U.: arbitrary units. (D) An arbitrary object "ON" imaged with subdiffraction resolution. Line width of the object is about 40 nm. The hyperlens is made of 16 layers of Ag/Al_2O_3. Reprinted with permission from "Zhaowei Liu, Hyesog Lee, Yi Xiong, Cheng Sun, Xiang Zhang, Science 315. 1686 (2007)" of copyright ©2007 of AAAS.

The object imaged with hyperlens was a pair of 35-nm-wide lines spaced 150 nm apart. Upon illumination, the scattered evanescent field from the object enters the anisotropic medium and propagates along the radial direction. Because of the conservation of angular momentum, the tangential wave vectors are progressively compressed as the waves travel outward, resulting in a magnified image at the outer boundary of the hyperlens. Hence the magnified image (350-nm spacing) can be clearly resolved with an optical microscopy.

3. Plasmonic lens on the basis of subwavelength metallic structures

3.1. Subwavelength metallic structure for superfocusing

3.1.1. One-dimensional structures for focusing

In this section, we presented two types of tuning methods for the purpose of phase modulation: depth tuning [23] and width tuning [24-26] approaches.

3.1.1.1. Depth-tuned. strctures

Three types of plasmonic slits (convex, concave, and flat/constant groove depth) with different stepped grooves have been designed and fabricated to achieve efficient plasmonic focusing and focal depth modulation of the transmitted beam. Figure 5 shows the fabricated depth-tuned plasmonic lens using focused ion beam milling [27]. The general design of the plasmonic slit is shown in Fig. 6 (a) [28]. When a TM polarized (magnetic field parallel to the y-direction) monochromatic plane wave impinges on the slit, it excites collective oscillations of the electrons at the surface, which is known as SPPs. The SPPs propagate along the surface of the metal film and are diffracted to the far-field by the periodic grooves, which are designed with a width smaller than half of the incident wavelength to allow a high diffraction efficiency [29]. Constructing interference of such diffracted beams leads to the focusing effect at a certain point on the beam axis [30]. Since the diffracted beams are modulated by the nanometric grooves, through adjusting the parameters of the grooves (such as our width, depth, period and number), the diffracted beams can be fully manipulated resulting in a tailored ultra-compact lens with subwavelength resolution and nanometer accuracy [31]. Most interestingly, it has been numerically found [32] that the relative phase at the exit end of the slit increases steadily with the increasing groove depth, making it possible to achieve continuous phase retardation by simply designing surrounding grooves with stepped depths as shown in Figs. 6 (b) and 6 (c). This has led to a great simplification of the plasmonic lens design without increasing the groove number or generating a bump on the metal film [33].

Figure 7 (a) presents a detailed comparison of the measured intensity distribution with simulation using FDTD at the slit cross section (along x-direction, as indicated by the white dot line in Fig. 2 (a) [23]). A good agreement has been found between the experiment and the theoretical prediction except that the measured full-width at a half-maximum (FWHM) of the central lobe (approximately 281 nm) is slightly larger than the calculated value of about 230 nm. This is because the measured intensity distribution approximately equals to the convolution of the finite probe size (30-80nm) and the actual intensity distribution of the transmitted light.

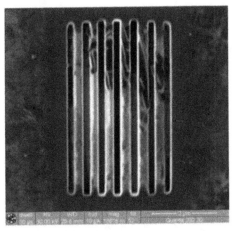

Figure 5. FIB image of the fabricated depth-tuned nanostructure (type of concave) on the Ag thin film with thickness of 200 nm.

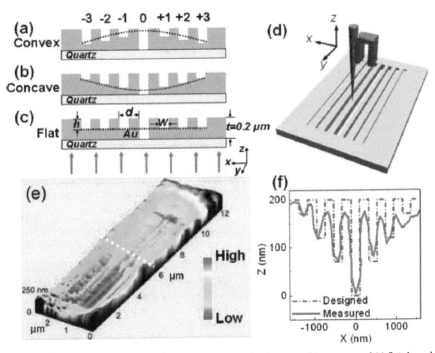

Figure 6. Schematic drawing of the nanoplasmonic slits with (a) convex, (b) concave and (c) flat shaped profiles. (d) Schematic drawing of near-field measurement setup. (e) Measured topographic image of the slit with concave corrugations. The marked area 'A' shows a larger overall depth than that of the area 'B'. (f) Cross section of the concave groove-slit at the position indicated by the dashed line in (e).

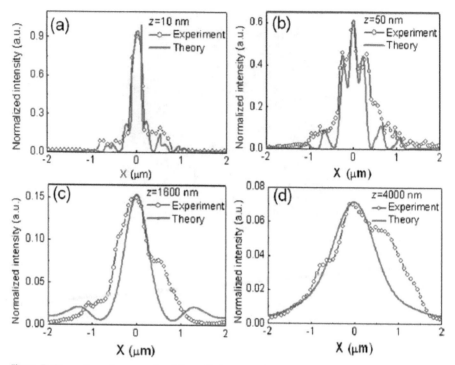

Figure 7. Comparison of measured and theoretical cross sections at x=0 in Fig. 2 of the reference paper for (a) z=10 nm, (b) z=50 nm (c) z=1600 nm and (d) z=4000 nm.

Near-field measurement reveals unambiguously the light interaction with the slits and confirms the functionalities of the nanoplasmonic lens. The simple plasmonic lens demonstrated in this paper can find broad applications in ultra-compact photonic chips particularly for biosensing and high-resolution imaging. Among the three types metallic structures, the type of convave structure has best focusing performance.

In adition, V-shaped influence on focusing performance was analyzed in fabrication point of view [34]. The incident angle dependance on the focusing properties was discussed also in Ref. [35].

Regarding fabrication of the metallic structures with depth-tuned grooves, it is worthy to point out that the geometrical characcrerization issue using atomic force microscope after focused ion beam direct milling [31]. Large measurement error is found during geometrical characterization of the nanostructures by use of an atomic force microscope (AFM) working in tapping mode. Apex wearing and 34° full cone angle of the probe generate the measurement errors during characterizing the nanostructures with the feature size of 200 nm and below. To solve this problem, a FIB trimmed AFM probe is employed in the geometrical characterization, as shown in Fig. 8. The results show that the error is improved greatly using the trimmed probe.

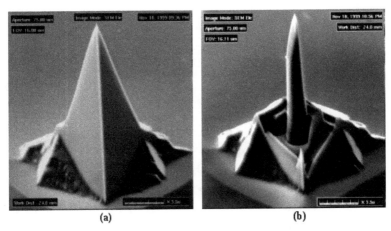

Figure 8. AFM probe for tapping mode. (a) the commercial probe with half cone angle of 17° and material of Si₃N₄. (b) the FIB trimmed probe with high aspect ratio.

Influence of polarization states on focusing properties of the depth-tuned metallic structures was reported [31]. The structure was designed with geometrical parameteres shown as Fig. 9. Figure 10 shows the total electric-field intensity $|E|^2 = |Ex|^2 + |Ey|^2 + |Ez|^2$ at x-y plane along the longitudinal direction at z = 1.35 μm at λ=420 nm for the (a) elliptical polarization (EP), (b) circular polarization (CP), and (c) radial polarization (RP) cases. The intensity of transverse electric field, $|Ex|^2 + |Ey|^2$, is significantly tuned. In the figure, the intensity along the horizontal (x) is equal to that along the vertical (y), while the intensity along the

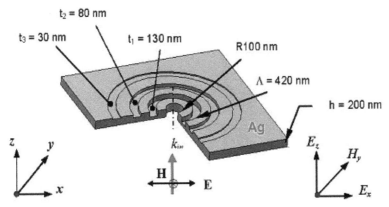

Figure 9. An annular plasmonic lens having a depth-tuned structure (groove depths, t₁= 130 nm, t₂= 80 nm and t₃=30 nm) milled in the output side of a Ag thin film (thickness, h = 200nm). Other structure parameters are: central hole diameter=200 nm, groove width=200 nm, and groove period=420 nm. The structure is incident with TM-polarized light having electric vector along the x direction. Reprinted with permission from "Jun Wang, Wei Zhou and Anand K. Asundi, Opt. Express 17, 8137-8143 (2009)." of copyright ©2009 Optical Society of American.

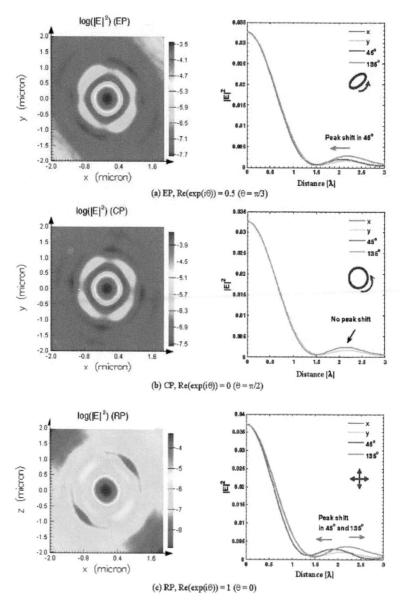

Figure 10. Total electric field (left) transmitted through the structure under illumination using different polarization states, including (a) CP, (b) EP, and (c) RP, showing the phase modulation effect on the beam profile (right) along the transverse direction in x and y and diagonal directions along 45° and 135° with respect to the x. Refer to Fig. 1 for the directions. Reprinted with permission from "Jun Wang, Wei Zhou and Anand K. Asundi, Opt. Express 17, 8137-8143 (2009)." of copyright ©2009 Optical Society of American.

diagonal directions (45° and 135° with respect to the x is tuned), the peak shift is observed at the side lobe, which is <0.1λ for the EP case. In Fig. 10 (a), in the direction of 45°. And the beam in 45° is narrower than that in the x- or y-directions. In addition, the phase function Re(exp(iθ)) indicates tuning capability. Re (exp(iθ)) = 0.5, where θ = π/3, for the EP case, and the phase function becomes 0 for the CP and 1 for the RP case. For example, in Fig. 10 (b), the uniformly distributed total-electric-field intensity is observed in the x-y plane, while, in Fig. 10 (c), the peak shifts 0.2λ in the 45° larger than that for the EP case, and much narrow beam is observed in the same direction. The same plasmonic modes are observed for CP, EP, or RP polarization cases as for TM case. Using a polarized plane wave the transverse electric field is tuned; the tuning 45° and 135° with respect to effect on focus spot is observed along the diagonal directions in 45° the x-direction. Of the cases, RP approach forms the smallest focus spot along the 45° using Re[exp(iθ)]=1, showing maximum tuning capability, while CP approach the phase function Re[exp(iθ)]=0 forms a symmetrically electric field distribution in the focal plane. Phase function indicates the tuning capability.

3.1.1.2. Width-tuned structures

A novel method is proposed to manipulate beam by modulating light phase through a metallic film with arrayed nano-slits, which have constant depth but variant widths. The slits transport electro-magnetic energy in the form of surface plasmon polaritons (SPPs) in nanometric waveguides and provide desired phase retardations of beam manipulating with variant phase propagation constant. Numerical simulation of an illustrative lens design example is performed through finite-difference time-domain (FDTD) method and shows agreement with theory analysis result. In addition, extraordinary optical transmission of SPPs through sub-wavelength metallic slits is observed in the simulation and helps to improve elements's energy using factor.

To illustrate the above idea of modulating phase, a metallic nano-slits lens is designed [24]. The parameters of the lens are as follows: D = 4 μm, f = 0.6 μm, λ= 0.65 μm, d=0.5 μm, where D is the diameter of the lens aperture, f the focus length, the wavelength and d the thickness of the film. The two sides of the lens is air. The schematic of lens is given in Fig. 11, where a metallic film is perforated with a great number of nano-slits with specifically designed widths and transmitted light from slits is modulated and converges in free space.

After numerous iterations of calculation using the FDTD algorithm, the resulting Poyinting vector is obtained and showed in Fig. 12 (a). A clear-cut focus appears about 0.6 micron away from the exit surface, which agrees with our design. The cross section of focus spot in x direction is given in Fig. 12 (b), indicating a full-width at half-maximum (FWHM) of 270 nm. The extraordinary light transmission effects of SPPs through sub-wavelength slits is also observed in the simulation with a transmission enhance factor of about 1.8 times. These advantages promise this method to find various potential applications in nano-scale beam shaping, integrate optics, date storage, and near-field imaging ect.

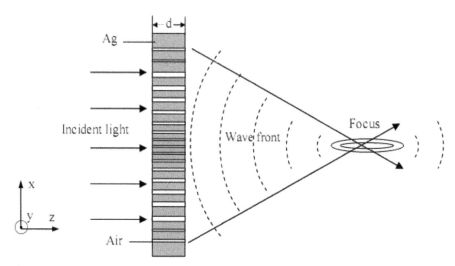

Figure 11. A schematic of a nano-slit array with different width formed on thin metallic film. Metal thickness in this configuration is d, and each slit width is determined for required phase distribution on the exit side, respectively. A TM-polarized plane wave (consists of Ex, Hy and Ez field component, and Hy component parallel to the y-axis) is incident to the slit array from the left side. Reprinted with permission from "H. F. Shi, C. T. Wang, C. L. Du, X. G. Luo, X. C. Dong, and H. T. Gao, Opt. Express 13, 6815-6820 (2005)." of copyright ©2009 Optical Society of American.

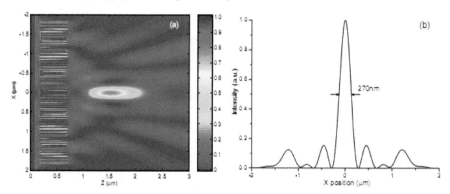

Figure 12. (a) FDTD calculated result of normalized Poynting Vector Sz for designed metallic nano-slits lens. Film thickness is 500nm, and the total slits number is 65. The structure's exit side is posited at z=0.7 μm. (b) Cross m.section of the focus at z=1.5 μm. Reprinted with permission from "H. F. Shi, C. T. Wang, C. L. Du, X. G. Luo, X. C. Dong, and H. T. Gao, Opt. Express 13, 6815-6820 (2005)." of copyright ©2009 Optical Society of American.

As an experimentlal verification example, Lieven et. al. [37] experimentally demonstrated planar lenses based on nanoscale slit arrays in a metallic film. The lens structures consist of optically thick gold films with micron-size arrays of closely spaced, nanoscale slits of varying widths milled using a focused ion beam. They found an excellent agreement between electromagnetic simulations of the design and confocal measurements on manufactured structures. They provide guidelines for lens design and show how actual lens behavior deviates from simple theory.

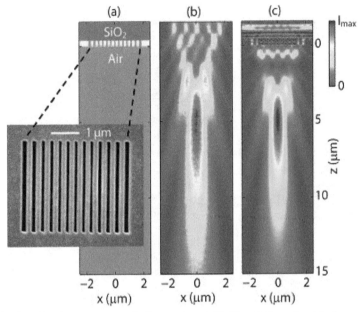

Figure 13. Planar lens based on nanoscale slit array in metallic film. (a) Geometry of the lens consisting of a 400nm optically thick gold film (yellow) with air slits of different widths (80 to150 nm) (light blue) milled therein on a fused silica substrate (dark blue). The inset shows a scanning electron micrograph of the structure as viewed from the air-side. (b) Focusing pattern measured by confocal scanning optical microscopy (CSOM). (c) Finite-difference and frequency-domain (FDFD) simulated focusing pattern of the field intensity through the center of the slits. In order to show the features of the focus spotclearly, the field intensity inside the slits is saturated. Reprinted with permission from "LievenVerslegers et. al., Nano Lett. 9, 235-238 (2009)" of copyright ©2009 Chemical Society of American.

The basic geometry consists of an array of nanoscale slits in an otherwise opaque metallic film (see Fig. 13 (a)). Figure13 shows the main results of the work, which combines fabrication, characterization, andsimulation. Pane l (a) shows the fabricated structure, while panels (b) and (c) represent the measured and simulated field intensity in across section through the center of the slits (along the x-direction). Both the measurement and the simulation clearly demonstrate focusing of the wave. The agreement between experiment and simulation is excellent. Moreover, the simulation image is generated using the designed parameters as the slit width rather than the actual slit width measured in the SEM, as is commonly done when comparing nanophotonics simulation and experiments. The agreemen there thus indicates the robustness in design and the fault tolerance of this approach for focusing. The effect of lens size can be exploited to control the focusing behavior as is shown clearly in Fig. 14. Both lenses introduce the same curvature to the incident plane wave as the lens from Fig. 13, since we consist of slits with the same width as the original design (2.5 μm long). By omitting one outer slit on each side for the lens in Fig. 14 (a), one gets the lens, as shown in Fig. 14 (b).

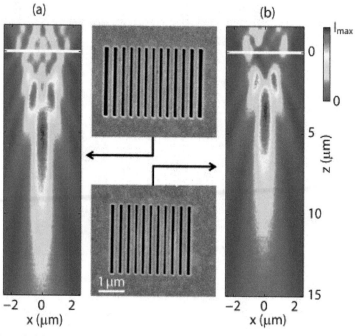

Figure 14. Control of the cylindrical lens behavior by design of nanoscale slit array parameters. Effect of lens size on focusing for (a) a lens with 13 slits (80-150nm by 2.5 μm) and (b) a lens with 11 slits (80-120nm by 2.5 μm). The white line gives an estimate of the lens position. Both scanning electron micrographs are on the same saclew. Reprinted with permission from "LievenVerslegers et. al., Nano Lett. 9, 235-238 (2009)" of copyright ©2009 Chemical Society of American.

This first experimental demonstration is a crucial step in the realization of this potentially important technology form any applications in optoelectronics. Moreover, the design principles presented here for the specil case of a lens can be applied to construct a wide range of optical components that rely on tailoring of the optical phase front.

3.1.2. Two-dimensional structures for focusing

3.1.2.1. Circular grating-based metallic structures for focusing

Recently, Jennifer et. al. reported the generation and focusing of surface plasmon polariton (SPP) waves from normally incident light on a planar circular grating (constant slits width and period) milled into a silver film [38]. The focusing mechanism is explained by using a simple coherent interference model of SPP generation on the circular grating by the incident field. Experimental results concur well with theoretical predictions and highlight the requirement for the phase matching of SPP sources in the grating to achieve the maximum enhancement of the SPP wave at the focal point. NSOM measurements show that the plasmonic lens achieves more than a 10-fold intensity enhancement over the intensity of a single ring of the in-plane field components at the focus when the grating design is tuned to the SPP wavelength.

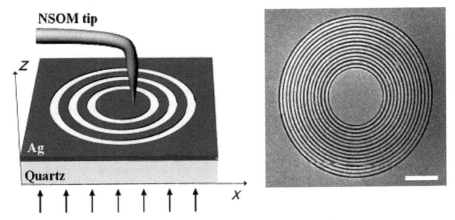

Figure 15. (a) Experimental scheme for near-field measurements. Circular gratings are cut into a silver film deposited on a quartz substrate. Laser light is normally incident from the quartz side, and the electromagnetic near-field is monitored with a metal coated NSOM tip. (b) SEM image of a sample with 15 rings. The scale bar is 5 microns. Reprinted with permission from "Jennifer M. Steele , Zhaowei Liu , Yuan Wang, and Xiang Zhang, Opt. Express 14, 5664-5670 (2006)." of copyright ©2006 Optical Society of American.

To investigate this focusing experimentally, rings with different periods were cut into 150 nm thick silver films. Silver was evaporated onto a quartz plate at a high rate to ensure a surface with minimal roughness. Rings were milled into the metal using an FEI Strata 201 XP focused ion beam (FIB), with the inner most ring having a diameter of 8 microns. Additional rings were added with a period either close to or far from resonance with the excited SPP waves. The surface plasmons were excited with linearly polarized laser light incident from the quartz side. The electromagnetic near-field of these structures was recorded using near-field scanning optical microscopy (NSOM) in collection mode using a metal coated NSOM tip. A metal coated tip was chosen over an uncoated tip to increase the resolution of the scan. Previous experimental results on samples with similar geometry compare favorably with computer simulations, indicating the interaction of the SPP near field with the metal tip is negligible. The measurement scheme can be seen in Fig. 15 (a) with an SEM image of a typical sample shown in Fig. 15 (b). The phase change of SPP waves across a barrier is an interesting issue that has received very little attention. If the slit width is much smaller than the SPP wavelength, the slit will have very little effect on the SPP and the phase change should be very small. However, if the slit width is on the order of the SPP wavelength, as the SPP waves cross a slit opposite charges will be induced on opposite sides of the slit, providing quasi-electrostatic coupling across the barrier. The authors calimed that it is possible that the phase change will be sensitive to the slit width. The number and period of rings, film material, and slit geometries provide experimental handles to tune the plasmonic lens to accommodate specific applications, making this technique a flexible plasmonic tool for sensing applications.

The following two sections below introduce the metallic subwavelength structures with chirped (variant periods) slits and nanopinholes acted as the plasmonic lenses for the purpose of superfocusing.

3.1.2.2. Illumination under linear polarization state

A novel structure called plasmonic micro-zone plate-like (PMZP) or plasmonic lens with chirped slits is put forth to realize superfocusing. It was proposed by Fu's group [39,40]. Unlike conventional Fresnel zone plate (CFZP), a plasmonic structure was used and combined with a CFZP. Configuration of the PMZP is an asymmetric structure with variant periods in which a thin film of Ag is sandwiched between air and glass. The PMZP is a device that a quartz substrate coated with Ag thin film which is embedded with a zone plate structure with the zone number N < 10. Figure 16 is an example of schematic diagram of the structure.

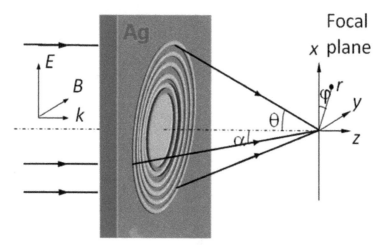

Figure 16. Schematic of the plasmonic micro-zone plate for super-focusing.

Following the electromagnetic focusing theory of Richard and Wolf (Richards & Wolf, 1959)[41], the electric field vector in the focal region is given by,

$$\mathbf{E}\,(r, z, \varphi) = -i\,[I_0\,(r, z) + I_2\,(r, z)\,\cos 2\varphi\,]\,\mathbf{i} - i\,I_2\,(r, z)\,\sin 2\varphi\,\mathbf{j} - 2I_1\,(r, z)\,\cos\varphi\,\mathbf{k}. \qquad (2)$$

Let **i, j, k** be the unit vectors in the direction of the co-ordinate axes. To be integral over the individual zones, I_0, I_1, I_2 are expressed as,

$$I_0(r,z) = \sum_n T_n \int_{\alpha_{n-1}}^{\alpha_n} \sqrt{\cos\theta}\,(1+\cos\theta) J_0(kr\sin\theta) \exp(ikz\cos\theta)\sin\theta d\theta,$$

$$I_1(r,z) = \sum_n T_n \int_{\alpha_{n-1}}^{\alpha_n} \sqrt{\cos\theta}\,\sin\theta J_1(kr\sin\theta) \exp(ikz\cos\theta)\sin\theta d\theta,$$

$$I_2(r,z) = \sum_n T_n \int_{\alpha_{n-1}}^{\alpha_n} \sqrt{\cos\theta}\,(1-\cos\theta) J_2(kr\sin\theta) \exp(ikz\cos\theta)\sin\theta d\theta.$$

The wave vector $k = 2\pi/\lambda$.

According to the equation, the intensity of lateral electric field component, $|E_x|^2$, follows the zero-order Bessel function J_0 of the first kind, while the intensity of longitudinal electric field component, $|E_z|^2$, follows the first-order Bessel function J_1 of the first kind. In the total electric field intensity distribution, all the field components add up. With high numerical aperture, this leads to not only asymmetry of the focus spot but also an enlarged focus spot. As shown in Figs. 17 (a)~(b), the total electric field and individual electric field components, $|E|^2$, $|E_x|^2$, & $|E_z|^2$, emerged from λ_{sp}-launched FZP lens are in comparison with that the total electric field and individual electric field components from a λ_{in}-launched PMZP lens. It is found that the intensity ratio, $|E_x|^2/|E_z|^2$, can increase up to 10 times for the λ_{sp}-launched PMZP lens. A focus spot having the polarization direction along the x direction is obtained.

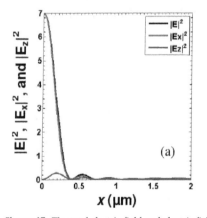 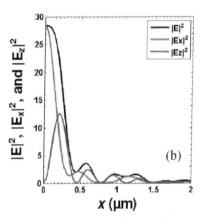

Figure 17. The total electric field and electric field components, $|E|^2$, $|E_x|^2$, and $|E_z|^2$, emerged from (a) a λ_{sp}-launched FZP lens and (b) a λ_{in}-launched FZP lens. It is observed that the proposed superlens enables to restrict the depolarization effect and produces a linearly-polarized focus spot having the polarization direction in the x direction.

The chirped slits can form a focal region in free space after the exit plane. The final intensity at the focal point is synthesized by iteration of each zone focusing and interference each other, and can be expressed as

$$I = \alpha \sum_{i=1}^{N} CI_0 \frac{4r_i}{\lambda_{SP}} e^{-(r_i/l_{SP})} \tag{3}$$

where I_0 is the incident intensity, r_i is the inner radius of each zone, i is the number of the zones, l_{SP} is the propagation length for the SPP wave, α is interference factor, and C is the coupling efficiency of the slits. C is a complicated function of the slit geometry and will likely have a different functional form when the slit width is much larger or much smaller than the incident wavelength

The PMZPs is an asymmetric structure. For an evanescent wave with given k_x, we have $k_{zj} = +[\varepsilon_j(\omega/c)^2 - k_x^2]^{1/2}$ for j=1 (air) and j=3 (glass) and $k_{zj} = +i[k_x^2 - \varepsilon_j(\omega/c)^2]^{1/2}$ for j=2 (Ag film). Superfocusing requires regenerating the evanescent waves. Thus the PMZP needs to be operated with the condition $|k_{z1}/\varepsilon_1 + k_{z2}/\varepsilon_2||k_{z2}/\varepsilon_2 + k_{z3}/\varepsilon_3| \to 0$. Physically, this would require exciting a surface plasmon at either the air or the glass side. For $E_\perp$ wave, a negative permittivity is sufficient for focusing evanescent waves if the metal film thickness and object are much smaller than the incident wavelength. Because electric permittivity $\varepsilon < 0$ occurs naturally in silver and other noble metals at visible wavelengths, a thin metallic film can act as an optical super lens. In the electrostatic limit, the p-polarized light, dependence on permeability μ is eliminated and only permittivity ε is relevant. In addition, diffraction and interference contribute to the transition from the evanescent waves to the propagation waves in the quasi-far-field region. Above all, the PMZPs form super focusing by interference of the localized SPP wave which is excited from the zones. This makes it possibly work at near and quasi-far-field with lateral resolution beyond diffraction limit. Also the PMZP has several zones only, its dimension is decreased greatly compared the CFZP.

As an example, an appropriate numerical computational analysis of a PMZP structure's electromagnetic field is carried out using finite-difference and time-damain (FDTD). It is illuminated by a plane wave with a 633 nm incident where Ag film has permittivity $\varepsilon_m = \varepsilon_m' + i\varepsilon_m'' = -17.6235 + 0.4204i$. An Ag film with thickness $h_{Ag} = 300nm$ centered at z=150 nm has an embedded micro-zone-plate structure. Zone number N=8, and outer diameter OD=11.93 μm. The widths of each zone from first ring to last ring, calculated by using the conventional zone plate equations, are 245, 155, 116, 93, 78, 67, 59, and 52 nm. In the FDTD simulations, the perfectly

matched layer boundary condition was applied at the grid boundaries. Figure 17 is the simulation result. From the result, the simulated focal length of the PMZP, f_{PMZP} and depth of focus (DOF) are larger than those of the designed values using the classical equations $r_n = (n\lambda f_{FZP} + n^2\lambda^2/4)^{1/2}$ and $DOF = \pm 2\Delta r^2/\lambda$, where n=1, 2, 3,..., f_{PMZP} is the designed principal focal length of Fresnel zone plates and given in terms of radius R of the inner ring

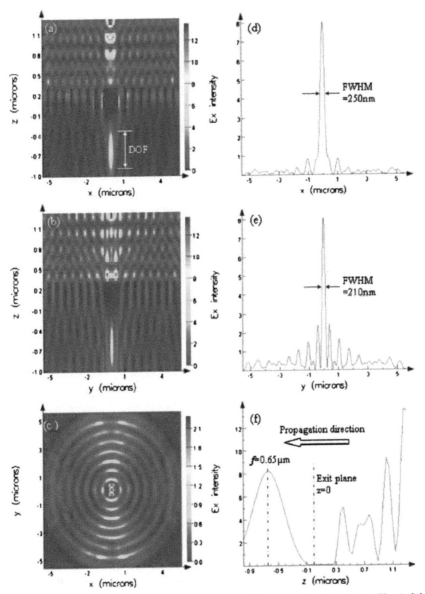

Figure 18. The simulation result of the example of the PMZP. The propagation direction is z. Electric field intensity $|E_x|^2$ at (a) y-z plane, (b) x-z plane, and (c) x-y plane. Electric field transmission in the line z=-0.65 μm (calculated focal plane) at (d) x-z plane, y=0; (e) y-z plane, x=0; and (f) y-z plane, x=0. The designed focal length and outmost zone width using scalar theory is f=1μm and 53 nm, respectively. The calculated DOF is ~700 nm (scalar theory designed value is 8.85 nm) the site z=0 is the exit plane of the Ag film.

and incident wavelength by $f_{FZP} = R^2 / \lambda$, Δr is the outmost zone width, and λ is the incident wavelength. It may be attributed to the SPPs wave coupling through the cavity mode and is involved for the contribution of the beam focusing. The focusing is formed by interference between the SPPs wave and the diffraction waves from the zones.

Figures 18 (a)~(f) are electric intensity distribution $|E_x|^2$ for $E_\perp$ wave in z-y, x-z, and x-y plane, respectively. The numerical computational analysis of the electromagnetic field is carried out using finite-difference and time-domain (FDTD) algorithm. It can be seen that focused spot size [full width at half maximum (FWHM)] at y-z plane is smaller than the one at x-y plane. For conventional Fresenl zone plate, when zone numbers are few, the first sidelobe is large. In contrast, our PMZP has much lower first sidelobe. Suppressed sidelobe at y-z plane is higher than the one at XOZ plane due to the incident wave with $E_\perp$ wave. Transmission with the $E_\perp$ wave illumination is extraordinarily enhanced due to the excited SPP wave coupling which is then converted to propagation wave by diffraction. Figure 2 (f) shows that there is only one peak transmission after exit plane of the Ag film. Furthermore, Fabry-Pérot-like phenomenon is found through the central aperture during the SPP wave coupling and propagation in cavity mode, as shown in Figs.17 (a), (b) and (f). It plays a positive role for the enhancement transmission.

Further characterization of the plasmonic lens was done,as shown in Figs. 19 (a)-(d) [42]. The Au thin film of 200 nm in thickness was coated on quart substrate using e-beam evaporation technique. The lens was fabricated using focused ion beam (FEI Quanta 200 3D dual beam system) direct milling technique, as shown in Fig. 19 (b). Geometrical characterization was performed using an atomic force microscope (Nanoscope 2000 from DI company). Figure 19 (c) shows topography of the FIB fabricated plasmonic lens. The optical measurement was performed with a near-field optical microscope (MultiView 2000[TS] from Nanonics Inc. in Israel) where a tapered single mode fiber probe, with an aperture diameter of 100 nm, was used working in collection mode. The fiber tip was raster scanned at a discrete constant height of 500 nm, 1.0 μm, 1.5 μm, 2.0 μm, 2.5 μm, 3.0 μm, 3.2 μm, 3.5 μm, 3.7 μm, 4 μm, 4.5 μm, and 5 μm, respectively, above the sample surface, and allowing us to map the optical intensity distribution over a grid of 256×256 points spanning an area of 20×20 μm². Working wavelength of the light source is 532 nm (Nd: YAG laser with power of 20 mW). Additionally, a typical lock-in amplifier and optical chopper were utilized to maximize the signal-to-noise ratio. Figure 19 (d) shows the measured three-dimensional (3D) electric field intensity distribution of the lens at propagation distance of 2.5 μm.

Figure 20 is a re-plotted 3D image of the NSOM measured intensity profiles along x-axis probed at the different propagation distance z ranging from 5 nm to 5 μm. It intuitively shows the intensity distribution along propagation distance. It can be seen that the peak intensity is significantly enhanced from 0.01 μm to 1 μm, and then degraded gradually in near-field region because of SPP-enhanced wave propagation on Au surface vanished in free space when z >1 μm. Only the interference-formed beam focusing region exits in near-field region. It is also in agreement with our calculated results. For more information, please see Ref. [43].

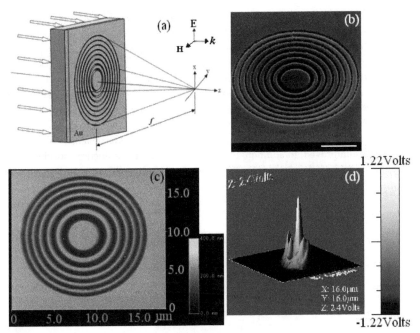

Figure 19. (a) Schematic diagram of the sandwiched plasmonic lens with chirped circular slits corrugated on Au film. Width of the outmost circular slit is 95 nm. Lens dimension (outer diameter) is 12 µm. (b) Scanning electron microscope image of the lens fabricated using focused ion beam milling technique. The scale bar is 4 µm. (c) AFM measurement result: topography of the fabricated lens. (d) NSOM characterization result of the lens: 2D E-field intensity distribution at propagation distance of 2.5 µm

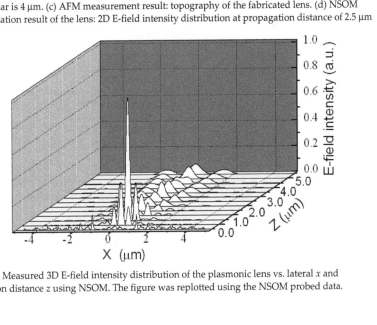

Figure 20. Measured 3D E-field intensity distribution of the plasmonic lens vs. lateral x and propagation distance z using NSOM. The figure was replotted using the NSOM probed data.

A hybrid Au-Ag subwavelength metallic zone plate-like structure was put forth for the purpose of preventing oxidation and sulfuration of Ag film, as well as realizing superfocusing, as shown in Fig. 21 [43]. The Au film acts as both a protector and modulator in the structure. Focusing performance is analyzed by means of three-dimensional (3D) finite-difference and time-domain (FDTD) algorithm-based computational numerical calculation. It can be tuned by varying thicknesses of both Au and Ag thin films. The calculated results show that thickness difference between the Au and Ag thin films plays an important role for transmission spectra. The ratio of Au to Ag film thicknesses, h_{Au}/h_{Ag}, is proportional to the relevant peak transmission intensity. In case of $h_{Au} \approx h_{Ag}$=50 nm, both transmission intensity and focusing performance are improved. In addition, the ratio h_{Au}/h_{Ag} strongly influences position of peak wavelengths λ_{Au} and λ_{Ag} generated from beaming through the metallic structures.

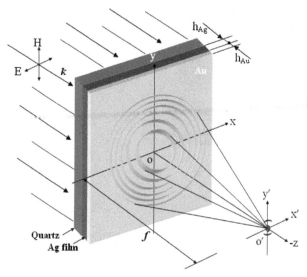

Figure 21. Schematic of the plasmonic micro-zone plate super-focusing with focal length f. It is illuminated by a plane wave with 633 nm incident wavelength. In our FDTD simulations the perfectly matched layer (PML) boundary condition was applied at the grid boundaries.

The following features were found from the calculation results:

1. Thickness difference between the Au and Ag thin films plays important role for transmission spectra. It determines peak transmission intensity. Ratio of film thicknesses $\alpha = h_{Au}/h_{Ag}$ is proportional to the relevant peak intensity. It may attributed to the resonant wavelength of the hybrid structure which is proportional to α.

2. In the case of $h_{Au} \approx h_{Ag} = 50$ nm, $\alpha \approx 1$, both transmission intensity and focusing performance are improved in comparison to the other cases (fixing $h_{Au} = 50$ nm and varying h_{Ag} from 10 nm to 200 nm).

3. For $h_{Ag} = 200$ nm, both transmission intensity and focusing performance are improved gradually with increasing h_{Au}. However, unlike the fixed 200 nm thickness of the Ag

film, fixing h_{Au} = 50 nm and 200 nm respectively and varying h_{Ag}, the corresponding optical performances are not improved gradually with increasing h_{Ag}.

4. Ratio $\alpha = h_{Au}/h_{Ag}$ strongly influences position of peak wavelengths λ_{Au} and λ_{Ag} generated from the metallic subwavelength structures.

The calculation results show that thickness of both the Au and Ag thin films has significant tailoring function due to the great contribution to superfocusing and transmission. Improved focusing performance and enhanced transmission can be obtained if h_{Au} and h_{Ag} match each other. This hybrid subwavelength structure has potential applications in data storage, nanophotolithography, nanometrology, and bio-imaging etc.

However, the rings-based structures have higher sidelobes. To supress the sidelobes, a circular holes-based plasmonic lens was reported, as shown in Fig. 22 [44]. In the plasmonic lens with fixed pinhole diameters, propagation waves still exist for much reduced periodicity of pinholes due to the SPPs wave coupling, which interferes with the diffraction wavelets from the pinholes to form a focusing region in free space. Increasing incident wavelength is equivalent to reducing the pinhole diameters, and rapid decay of the EM field intensity will occur accordingly. The superlens proposed by the authors has the advantages of possessing micron scale focal length and large depth of focus along the propagation direction. It should be especially noted that the structure of the superlens can be easily fabricated using the current nanofabrication techniques, e.g. focused ion beam milling and e-beam lithography.

To further improve focusing quality of the circular holes-based plasmonic lens, an elliptical nanoholes-based plasmonic lens was put forth, as shown in Fig. 9 [45]. The plasmonic lens is

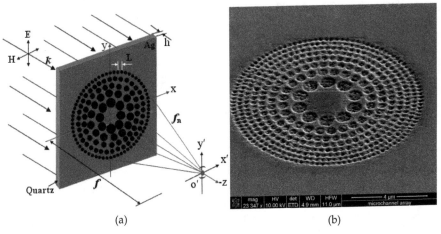

(a)   (b)

Figure 22. (a) Schematic of the pinhole array with focal length f. Lateral central distance L determines wave coupling between the neighboring holes. The pinholes are uniformly distributed in the zones. It is illuminated by a plane wave with 633 nm incident wavelength and p-polarization (transverse-magnetic field with components of E_x, H_y, and E_z). The perfectly matched layer (PML) boundary condition was applied at the grid boundaries in the three-dimensional FDTD simulation. (b) Nanofabrication of the lens using focused ion beam.

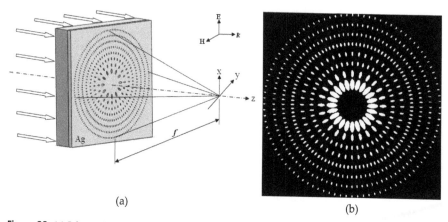

(a)   (b)

Figure 23. (a) Schematization of the pinhole array with focal length f. Lateral central distance L determines of wave coupling between the neighbored holes. The pinholes are uniformly distributed along the zones. It is illuminated by various waves with 633 nm incident wavelength. And we have different polarization states such as TM, EP and RP. (b) layout of elliptical pinholes with total 8 rings $\delta = 0.6$ being used in our computational numerical calculation.

composed of elliptical pinholes with different sizes distributed in different rings with variant periods. Long-axis of the ellipse is defined as $a_n=3w_n$, whereas w_n is width of the corresponding ring width, and n is the number of rings. A thin film of Ag coated on the glass substrate is perforated by the pinholes. The numbers of w_n and radius for different rings are listed in Table1.

Ring No.	1	3	3	4	5	6	7	8
Ring radius(μm)	1.41	2.11	2.78	3.43	4.08	4.73	5.37	6.01
w_n (nm)	245	155	116	93	78	67	59	53

Table 1. The numbers of w_n for different rings with orders from inner to outer (designed f=1μm)

As an example, the authors studied the case of 200 nm thickness Ag film coated on quartz substrate and designed a nanostructure with 8 rings on the metal film (see Fig. 23). The pinholes are completely penetrated through the Ag film. The number of pinholes from inner to outer rings is 8, 20, 36, 55, 70, 96, 107, and 140, respectively. Outer diameter of the ring is 12.05 μm. Radius of the rings can be calculated by the formula $r_n^2 = 2nf\lambda + n^2\lambda^2$, where f is the focal length for working wavelength of λ=633 nm which we used in our works. For simplicity, we define a ratio of short-axis to long-axis $\delta= b/a$ (where a is the length of long-axis of the elliptical pinholes, and b the short-axis). The used metal here is Ag with dielectric constant of $\varepsilon_m =-17.24 + i0.498$ at λ=633 nm, and ε_d=1.243 for glass. The incident angle θ is 0° (normal incidence). In our analysis, we simulated the cases of the ratios: δ = 0.1, 0.2, and 0.4, respectively. The ultra-enhanced lasing effect disappears when the ratio $\delta\rightarrow1$ (circular pinholes). Orientation of the pinholes is along radial direction. The pinholes symmetrically distribute in different rings with variant periods. It can generate ultra-enhanced lasing effect and realize a long focal length in free space accordingly with extraordinarily elongated depth of focus (DOF) of as long as 13 μm under illumination of plane wave in linear y-polarization.

Figure 24 shows our computational results: E-field intensity distribution $|Ey|^2$ at x-z and y-z planes for (a) and (b) δ=0.1; (c) and (d) δ=0.2; and (e) and (f) δ=0.4, respectively. It can be seen that E-field intensity distribution is symmetric due to linear y-polarization which is formed by uniformly rotating linear polarization radiating along radial directions. By means of interfering constructively, and the Ez component interferes destructively and vanishes at the focus. Thus we have $|E|^2=|Ex|^2+|Ey|^2+|Ez|^2=|Ey|^2$, whereas Ex=0, and Ez=0. The lasing effect-induced ultra-long DOF is 7 μm, 12 μm, and 13 μm for δ=0.1, 0.2, and 0.4, respectively. It is three orders of magnitude in comparison to that of the conventional microlenses. The ultra-enhanced lasing effect may attribute to the surface plasmon (SP) wave coupling in the micro- and nano-cavity which form Fabry-Pérot resonance while the beam passing through the constructive pinholes. Calculated full-width and half-maximum (FWHM) at propagation distance z= -7 μm in free space is 330 nm, 510 nm, and 526 nm for δ=0.1, 0.2, and 0.4, respectively.

Lasing effect of the plasmonic lens with extraordinarily elongated DOF has the following unique features in practical applications:

1. In bioimaging systems such as confocal optical microscope, three-dimensional (3D) image of cells or molecular is possible to be obtained and conventional multilayer focal plane scanning is unnecessary.
2. In online optical metrology systems, feed-back control system can be omitted because height of surface topography of the measured samples as large as ten micron is still within the extraordinarily elongated DOF of the plasmonic lenses.
3. In plasmonic structures-based photolithography systems, the reported experimental results were obtained at near-field with tens nanometers gap between the structure and substrate surface. Apparently, it is difficult to control the gap in practical operation. However, using this plasmonic lens, the working distance between the structure and substrate surface can be as long as 12 μm and even longer. Practical control and operation process will be much easier and simplified than the approaches before in case of using this lens.

Like the discussion above regarding illumination with different polarization states, influence of polarization states on focusing properties of the the plasmonic lenses with both chirped circular slits and elliptical nanopinholes were reported [46, 47].

The lens was fabricated using focused ion beam directly milling technique, and characterization of the elliptical nanopinholes-based plasmonic lens was carried out using NSOM, as shown in Figs. 25 (a) and (b). Currently, this work is still in progress in the research group.

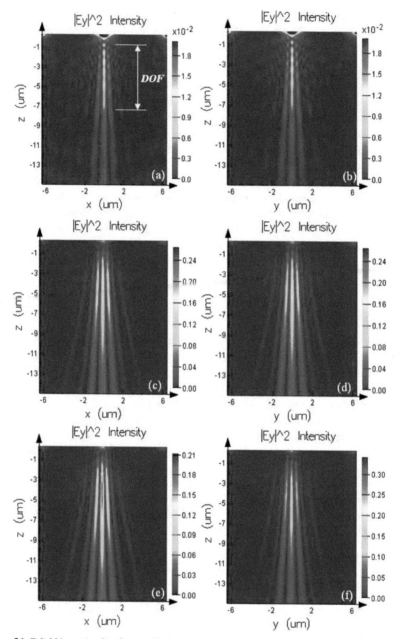

Figure 24. E-field intensity distribution $|E_y|^2$ at x-z and y-z planes for (a) and (b) δ=0.1; (c) and (d) δ=0.2; and (e) and (f) δ=0.4, respectively.

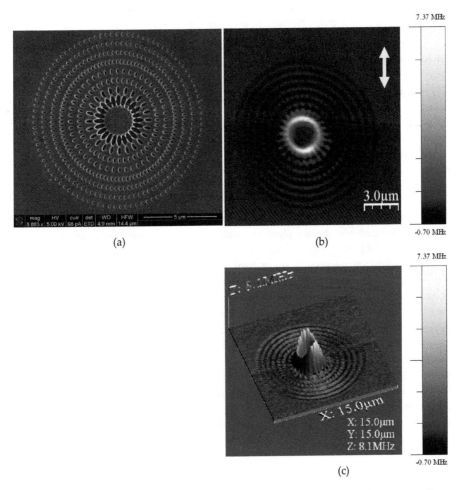

Figure 25. (a) SEM micrograph of the elliptical nanopinholes-based plasmonic lens. (b) 2D image of NSOM probing along propagation distance at z=20nm in free space. The arrow indicates direction of linear polarization. (c) 3D image of NSOM probing.

Here we addressed an issue here from fabrication point of view. Real part of Au permittivity ε_{Au} will be increased due to Ga+ implantation of the FIB directly etching process. Theoretically, propagation constant k_{SP} will be increased due to the large real part of dielectric constant as $k_{SP} = \dfrac{\omega}{c}\sqrt{\dfrac{\varepsilon_m \varepsilon_d}{\varepsilon_m + \varepsilon_d}}$, whereas ω is the incident frequency, c is the speed of light in vacuum, ε_m and ε_d is dielectric constant of metal and dielectric, respectively. The increased k_{SP} will cause strong transmission enhancement and generate extended skin depth in free space which is helpful for formation of the focusing region and makes a positive contribution on the plasmonic focusing accordingly.

An elliptical nano-pinholes-based plasmonic lens was studied experimentally by means of FIB nanofabrication, AFM imaging, and NSOM characterization for the purpose of proof of plasmonic finely focusing. Both modes of sample scan and tip scan were employed for the lens probing. For the NSOM-based optical characterization of the plasmonic lenses, both of them have their own characteristics. The former can generate a bright-field like image with strong and uniform illumination; and the latter can produce a dark-field like image with high contrast which is helpful for checking focusing performance of the lenses. Our experimental results demonstrated that the lens is capable of realizing a subwavelength focusing with elongated depth of focus.

3.1.2.3. Illumination under radial polarization state

Most recently, the circular rings-based plasmonic lens was experimentally demonstrated [48, 49]. The focusing of surface plasmon polaritons by a plasmonic lens illuminated with radially polarized light was investigated. The field distribution is characterized by near-field scanning optical microscope. A sharp focal spot corresponding to a zero order Bessel function is observed. For comparison, the plasmonic lens is also measured with linearly polarized light illumination, resulting in two separated lobes. Finally, the authors verify that the focal spot maintain sits width along the optical axis of the plasmonic lens. The results demonstrate the advantage of using radially polarized light for nanofocusing applications involving surface plasmon polaritons. Figures 26 (a) and (b) are reported NSOM characterization results. For comparison purposes, this theoretical cross section is also shownin Figure 26 (b). The profile of the Bessel function can be clearly observed. Neglecting the contribution of the E component, the theoretical spot size (based on full width half-maximum criterion) is 380nm. The measured spot size is slightly larger, 410 ± 39nm (error was estimated by taking several cross sections through the center of the PL along different directions).

3.2. Subwavelength metallic structure for imaging

A planar lens based on nanoscale slit arrays in a metallic film is present here for subwavelength imaging in the far field. Figure 27 is the schematic of the optical imaging with metallic slab lens [51]. To illustrate the design, both object distance a and image distance b are set to be 1 μm. The aperture and the thickness of silver slab lens are 3μm and 300 nm respectively. The permittivity ε_m=-29.26+i1.348 is used for silver at 810 nm.

When the light with magnetic field polarized in the y direction impinged on the surface of silver slab, SPPs can be excited at the slit entrance. The SPPs propagate inside the slits in the specific waveguide modes until reaching the export where we radiate into free space and form the optical image at the desired position. For the imaging of object localized on the axis x=0, the phase retardation of light transmitted through the lens is giver by

$$\Delta\varphi(x) = 2n\pi + \Delta\varphi(0) + \frac{2\pi}{\lambda}(a + b - \sqrt{a^2 + x^2} - \sqrt{b^2 + x^2}) \qquad (4)$$

where n is an integer number. Therefore, the key point of designing the metallic lens is to determine the width and position of slits for appropriate phase retardation.

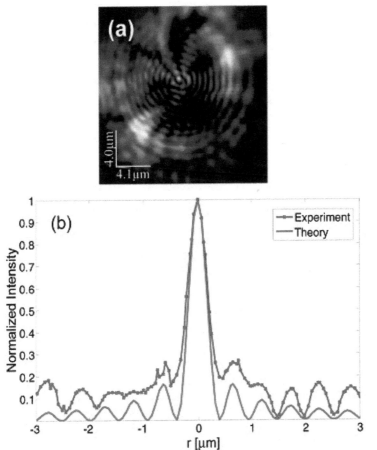

Figure 26. NSOM measurement showing SPP focusing in the plasmonic lens illuminated by radially polarized light. The NSOM probe was at a constant height of 2 μm above the PL. (a) 2D NSOM scan. Bright regions correspond to high intensity. (b) Normalized experimental (blue, cross markers) and theoretical (red, solid line) cross sections through the center of the PL. The sharp focus can be clearly observed. Reprinted with permission from "Gilad M.Lerman, Avner Yanai,and Uriel Levy, Nano Lett. 9, 2139-2143 (2009)." of copyright ©2009 Chemical Society of American.

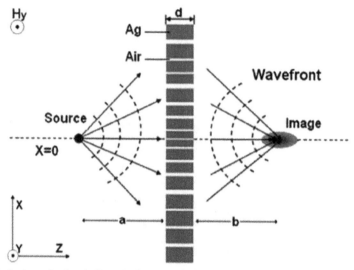

Figure 27. A schematic of optical imaging by a metallic slab lens with nanoslits. Reprinted with permission from "Ting Xu, et. al. Appl. Phys. Lett. 91,201501 (2007)" of copyright ©2007 American Institute of Physics.

Because of the complexity of the accurate description of the processes of SPPs' excitation, propagation, and coupling in metallic slits, the convenient and effective way is the approximation of these processed and making sure of good accuracy simultaneously. The coupling of SPPs during the propagation in slits is neglected, provided that the metallic wall between any two adjacent slits is larger than the skin depth in metal, about 24 nm for silver at a wavelength of 810 nm giver by

$$Z = \frac{1}{k_0}\left[\frac{\mathrm{Re}(\varepsilon_m)+\varepsilon_d}{\mathrm{Re}(\varepsilon_m)^2}\right]^{1/2}$$

(5)

The coupling effect occurred at the exit surface from neighboring slits is also omitted, compared with the intensity of directly radiated light from slits. The design above also displays that only a slight deviation is produced by coupling effect and it affects the image profile insignificantly. Assuming that the slit width is much smaller than a wavelength, it is justified to only consider the fundamental mode in the slit. Its complex propagation constant β in the slit is determined by the following equation:

$$\tanh(\sqrt{\beta^2 - k_0^2\varepsilon_d}\,w/2) = \frac{-\varepsilon_d\sqrt{\beta^2 - k_0^2\varepsilon_m}}{\varepsilon_m\sqrt{\beta^2 - k_0^2\varepsilon_d}}$$

(6)

where k_0 is the wave vector of light in free space, ε_m and ε_d are the permittivity of the metal and dielectric material inside the slits, and w is the slit width. The real and imaginary parts of β determine the phase velocity and the propagation loss of SPPs inside the metallic slit, respectively.

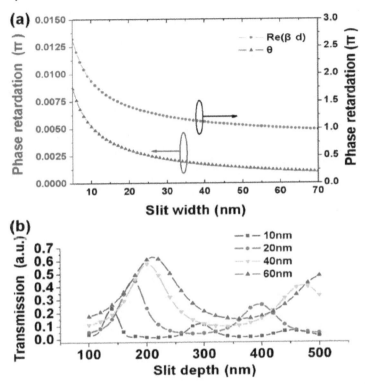

Figure 28. (a) Dependence of phase retardation on the slit width. Red and blue tags represent the contributions for phase retardation from the real part of propagation constant and multiple reflections. The wavelength of incident light is 810 nm. (b) Transmittance of optical field (810 nm wavelength) through a nanoslit vs the slit depth ranging from 100 to 500 nm. Slit widths are 10, 20, 40, and 60 nm. Reprinted with permission from "Ting Xu, et. al. Appl. Phys. Lett. 91,201501 (2007)" of copyright ©2007 American Institute of Physics.

To illustrate the validity of metallic slab lens, finite-difference time-domain simulations are performed. Figure 28 (a) illuminates the dependence of phase retardation on slit width. Figure 28 (b) plots the finite-difference time-domain (FDTD) simulated transmittance of light through a slit with variant thickness as ranging from 100 to 500 nm by normal incidence at the wavelength of 810 nm. Calculated steady optical field (the magnetic field $|Hy|^2$) of the simulation result is shown in Fig. 29. Obvious image spot can be seen at the position around Z=2.43 μm with the full width at half maximum of 396 nm approximately half of the incident wavelength. The slight focal shift of about 60 nm. Above all, the metallic slab lens displays a considerably good performance for imaging objects in the far field region.

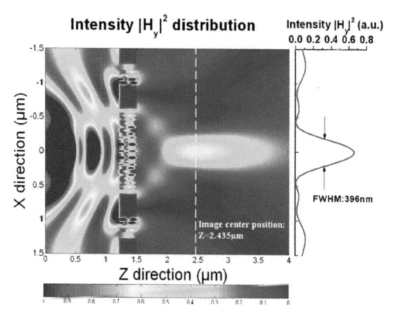

Figure 29. Calculated steady magnetic field intensity $|Hy|^2$ of the simulation results using FDTD method. The spirce is localized at X=0 and Z=0.2 μm. The metallic slab lens ranged from Z=1.2 μm to Z=1.5 μm. The radiated light from the source is TM polarized with a wavelength of 810nm. The curve at the right side represents the cross section of image plane at Z=2.435 μm. Reprinted with permission from "Ting Xu, et. al. Appl. Phys. Lett. 91,201501 (2007)" of copyright ©2007 American Institute of Physics.

In addition, nanorod array was reported being used for imaging [51]. Mark et. al reported parallel conducting wires as a lens for subwavelength microwave imaging [52].

4. Plasmonic Lens on the basis of waveguide modes

Another structure that can realize nanofocusing was theoretically reported [53]. SPPs propagating toward the tip of a tapered plasmonic waveguide are slowed down and asymptotically stopped when we tend to the tip, never actually reaching it (the travel time to the tip is logarithmically divergent). This phenomenon causes accumulation of energy and giant local fields at the tip. Focusing of fundamental cylindrical SPP wave is formed at apex of the taper tip, as shown in Fig. 30. Figure 31 displays the amplitudes of the local optical fields in the cross section of the system for the normal and longitudinal (with respect to the axis) components of the optical electric field. As SPP's move toward the tip, the SPP fields start to localize at the metal surface, and simultaneously, our wavelength is progressively reducing and amplitude growing. The field magnitudes grow significantly at small $|Z|$. The transverse x component grows by an order of magnitude as the SPP's approach the tip of the guide, while the longitudinal z component, which is very small far from the tip, grows relatively much stronger. The 3D energy concentration occurs at the tip of a smoothly tapered metal nanoplasmonic waveguide. This causes the local field increase by 3 orders of magnitude in intensity and four orders in energy density.

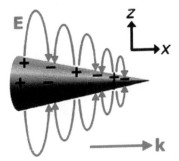

Figure 30. Geometry of the nanoplasmonic waveguide. The radius of the waveguide gradually decreases from 50 nm to 2 nm. Reprinted with permission from "Mark I. Stockman, Phys. Rev. Lett. 93, 137404 (2004)" with copyright © 2004 of American Society of Physics.

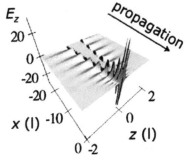

Figure 31. Snapshot of instanteneous fields (at some arbitrary moment $t=0$): longitudinal component Ez of the local optical electric field are shown in the longitudinal cross section (xz) plane of the system. The fields are in the units of the far-zone (excitation) field. Reprinted with permission from "Mark I. Stockman, Phys. Rev. Lett. 93, 137404 (2004)" with copyright © 2004 of American Society of Physics.

Most recently, Ewold *et. al.* reported our tapered waveguid structure for nanofocusing, as shown in Fig. 32 [54]. It was used for focusing of surface plasmon polaritons (SPPs) excited with 1.5 μm light in a tapered Au waveguide on a planar dielectric substrate by experiments and simulations. We find that nanofocusing can be obtained when the asymmetric bound mode at the substrate side of the metal film is excited. The propagation and concentration of this mode to the tip is demonstrated. No sign of a cutoff waveguide width is observed as the SPPs propagate along the tapered waveguide. Simulations show that such concentrating behavior is not possible for excitation of the mode at the low-index side of the film. The mode that enables the focusing exhibits a strong resemblance to the asymmetric mode responsible for focusing in conical waveguides. This work demonstrates a practical implementation of plasmonic nanofocusing on a planar substrate.

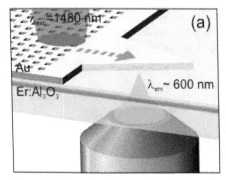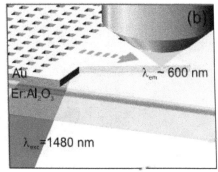

Figure 32. Schematic of the experimental geometry in the case of upconversion luminescence detection through the substrate (a), or from the air side of the sample (b). In both cases the SPPs are excited with infrared light at the Au/Al₂O₃ interface in the direction of the arrow. The red line schematically indicates the Er depth profile. Reprinted with permission from "Ewold Verhagen, Albert Polman, and L. (Kobus) Kuipers, Opt. Express 16, 45-57 (2008)." of copyright ©2008 Optical Society of American.

A fiber-pigtailed 1.48 µm diode pump laser is used as excitation source. Figure 32 shows a schematic of the hole array/taper geometry. The pitch of the hole array is chosen such that p-polarized light with a wavelength of 1.48 µm is diffracted to generate SPPs propagatin at the substrate side of the film. To maximize the excitation of the desired SPP mode, the excitation beam is focused to a 10 µm wide spot near the edge of the array. The triangularl shaped tapered waveguide starts at a distance of 6 µm from the edge of the excitation arra and has a base width of 12 µm and a length of 60 µm (taper angle 11°).

5. Plasmonic lens on the basis of curved chains of nanoparticles

Focusing of surface plasmon polaritons (SPPs) beams with parabolic chains of gold nanoparticles fabricated on thin gold films was reported [55]. SPP focusing with different parabolic chains is investigated in the wavelength range of 700-860nm, both experimentally and thoretically. Mapping of SPP fields is accomplished by making use of leakage radiation microscopy, demonstrating robust and efficient SPP focusing into submicron spots. Numerical simulations based on the Green's tenor formalism show very good agreement with the experimental results, suggesting the usage of elliptical corrections for parabolic structures to improve our focusing of slightly divergent SPP beams.

Shortly after the above work, excitation,focusing and directing of surface plasmon polaritons (SPPs) with curved chains of nanoparticles located on a metal surface is investigated both experimentally and theoretically by Evlyukhin et. al., as shown in Fig. 33 [56]. We demonstrate that, by using a relatively narrow laser beam (at normal incidence) interacting only with a portion of a curved chain of nanoparticles, one can excite an SPPs beam whose divergence and propagation direction are dictated by the incident light spot size and its position along the chain. It is also found that the SPPs focusing regime is strongly influenced by the chain inter-particle distance. Extensive numerical simulations of

Plasmonic Lenses

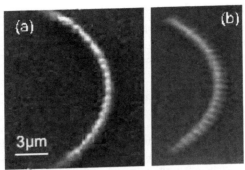

Figure 33. Scanning electron microscope (SEM) (a) the top image of the structure, (b) the image of the structure obtained with view angle 45°. The radius of curved chains of nanoparticles is equal to 10μm. The particle in-plane size (diameter) and inter-particle distance are estimated to be about 350nm and 850nm, respectively, the particle height is 300nm. Reprinted with permission from "A. B. Evlyukhin, et. al, Opt. Exp. 15, 16667-16680 (2007)." of copyright ©2007 Optical Society of American.

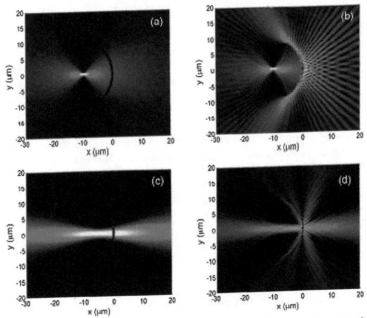

Figure 34. Magnitude of scattered electric field calculated above the gold surface with a curved chain (with R=10 μm and β=60°) of spheroid gold nanoparticles illuminated by a light beam at the wavelength of 800nm being incident perpendicular to the gold surface and polarized along x-direction. The waist W of the incident beam and the inter-particle (center-to-center) spacing D in the chain are (a) W=10μm, D=400nm; (b)W=10μm, D=800nm; (c)W =1.5 μm, D=400nm; (d)W =1.5 μm, D=800nm. Reprinted with permission from "A. B. Evlyukhin, et. al, Opt. Exp. 15, 16667-16680 (2007)." of copyright ©2007 Optical Society of American.

the configuration investigated experimentally are carried out for a wide set of system parameters by making use of the Green's tensor formalism and dipole approximation. Comparison of numerical results with experimental data shows good agreement with respect to the observed features in SPP focusing and directing, providing the guidelines for aproper choice of the system parameters.

When the inter-particle distance is smaller than the light wavelength, the pattern of the field magnitude distribution is relatively smooth [Fig. 34 (a)]. In this case, the illuminated part of the chain exhibits scattering properties that are similar to those of a continuous ridge. Note that straight ridges are frequently used for excitation of a divergent SPP beam on a metal surface in SPP experiments. When increasing the inter-particle distance, the individual particles of the chain become relatively independent sources of the scattered waves whose phases differ considerably, resulting in a complex interference pattern [Fig. 34 (b)] – a system of divergent SPPs rays. A similar trend is also seen for a relatively narrow incident light beam [see Figs. 34 (c) and (d)]. If the light spot size being determined by W is sufficiently small in comparison with the chain curvature radius R so that the diffraction angle of a SPPs beam is approximately equal to the focusing angle W/R, the focusing effect decreases and the maximum of SPPs intensity moves toward the nanoparticle chain. Strong SPPs focusing effects have been obtained for relatively lager W/R.

Apart from the chain particle-based plasmonic structures for nanofocusing, the reverse pattern: nanoholes, constructed in curved chains can also realize focusing of SPP wave, as shown in Fig. 35 [57]. The focused SPPs can be directly coupled into a waveguide located at the focal plane. The constructive interference of SPPs launched by nanometric holes allows us to focus SPP into a spot of high near-field intensity having subwavelength width. Near-field scanning optical microscopy is used to map the local SPP intensity. The resulting SPPs patterns and our polarization dependence are accurately described in model calculations based on a dipolar model for the SPP emission at each hole. Furthermore, we show that the high SPPs intensity in the focal spot can be launched and propagated on a Ag strip guide with a 250×50 nm^2 cross section, thus overcoming the diffraction limit of conventional optics. The combination of focusing arrays and nano-waveguides may serve as a basic element in planar nano-photonic circuits.

Not only focusing and imaging, the similar plasmonic structures can be used as beam sppliters [58] and beam shaping [59]. Nanofabrication of the plasmonic lenses were reported in Ref. [60-62].

6. Plasmonic Talbot effect of nanolenses

Previous introduction in Section 3 shows a common phenomenon which there several focal points exist along central axis in free space after exit plane of the nanolenses.

To explore its physical mechanism, we compared the phenomenon to another well-known story: Talbot effect, *i.e.* self-imaging. The self-imaging means that when a one-dimensional (1D) periodic structure is illuminated by the monochromatic plane wave, the image of that structure can be observed at the periodical distance from the back side of the structure. It is

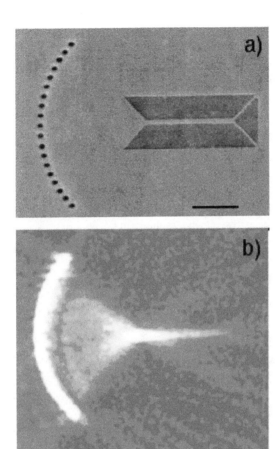

Figure 35. (a) SEM image of the focusing array coupled to a 250-nm-wide Ag strip guide; light gray, Ag; dark gray, Cr; scale bar, 2 μm.(b) NSOM image of the SPPs intensity showing focusing and guiding. Reprinted with permission from "Leilei Yin, et. al., Nano Lett. 2005, 5, 1399–1402." of copyright ©2005 Chemical Society of American.

a type of imaging by diffraction rather than an ordinary imaging of a lens. Actually, some researchers are still interested in the self-imaging effect since H. F. Talbot discovered it in 1836. And intensive theoretical and experimental studies regarding Talbot effect have been done since then. For example, the plasmon analogue of the self-imaging Talbot effect of a row of holes drilled in a metal film was described and theoretically analyzed, and suggested the potential applications in sensing, imaging, and optical interconnects on the basis of plasmon focal spots aimed at plasmon waveguides [63]. Subsequently, the Talbot effect regarding SPPs imaging on the basis of a rather different system by a quite different approach was studied theoretically and experimentally [62-64]. Furthermore, the Talbot effect for volume electromagnetic waves has been used in a variety of applications. And it is expected that the analogue for SPPs will be found applications in numerous nanoscale plasmonic devices.

Here, as an example, the Talbot effect of an Ag nanolens with five periodic concentric through the rings illuminated by a radially polarized light was computationally studied. Rigorous finite-difference and time-domain (FDTD) algorithm was employed in the computational numerical calculation. The results indicate that several focal points can be obtained at different locations due to the SPPs-related Talbot effect at $\lambda_{inc.}=248$ nm. The positions are quite different from that of values calculated by the Talbot distance equation reported in Ref. [65]. A minimum diameter of 100 nm at site of full width and half maximum (FWHM) was derived at the propagation distance of Z=396 nm. To further study the phenomenon in physics, it was compared with the traditional Talbot distance calculated using the scalar diffraction theory in the sections below.

For simplification of the analyses, the structure is simplified that is equivalent to a one-dimensional grating structure with the same geometrical parameters as the nanolens mentioned before due to its symmetry, as shown in Fig. 36.

The scalar field immediately behind an infinite grating at the original position (Z = 0) when it is illuminated by an unit intensity plane wave can be described by a Fourier series representing a weighted set of plane wave components as:

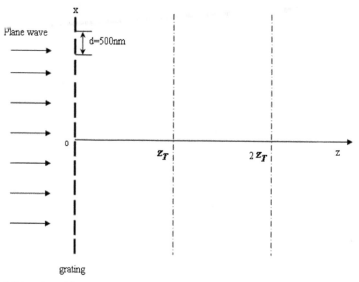

Figure 36. Self-imaging of the grating structure with the same geometrical parameters as the nanolens mentioned before.

$$U(x) = \sum_{n=-\infty}^{\infty} c_n \exp\left(j2\pi\frac{n}{d}x \right) \qquad (7)$$

where d is the grating period (is 500 nm here) and c_n is the nth Fourier coefficient. The coefficients represent the complex intensity and the phase.

According to the frequency domain analysis method, the diffraction field distribution U_z at a free-space propagation distance Z is given by

$$U_z(x) = \sum_{n=-\infty}^{\infty} c_n \exp\left(j2\pi\frac{n}{d}x\right)\exp\left[-j\pi\lambda z\left(\frac{n}{d}\right)^2\right]\exp(jkz) \tag{8}$$

Note that when

$$z = \frac{2md^2}{\lambda} \qquad (m = 1,2,3,...) \tag{9}$$

then

$$\exp\left[-j\pi\lambda z\left(\frac{n}{d}\right)^2\right] = 1 \tag{10}$$

U_z can be written as

$$U_z(x) = \sum_{n=-\infty}^{\infty} c_n \exp\left(j2\pi\frac{n}{d}x\right)\exp(jkz) = U(x)\exp(jkz) \tag{11}$$

Intensity distribution equals to the original intensity

$$I = |U_z(x)|^2 = |U(x)|^2 \tag{12}$$

The Talbot distance of a grating with a period of d is

$$z_T = \frac{2d^2}{\lambda} \tag{13}$$

Substituting d and λ with their corresponding values of 500 nm and 248 nm, respectively, we can get the Talbot distance of z_T=2.016 µm. It indicates that self-imaging of the periodic grating structures can be observed at positions of Z=2.016 µm, 4.032 µm, and 6.048 µm, etc., which is integral number times of z_T.

It was found that the result analyzed above by the scalar diffraction theory is quite different from the results calculated by the FDTD numerical analysis method. The former is far field diffraction, but the latter is SPPs coupling and interfering at near field. Thus the focal points do not repeat at the same positions as that of the scalar one along the propagation direction. The phenomenon described before can be attributed to that of the surface plasmons (SP). SPPs is excited at all azimuthal directions that interferes each other constructively and creates a strongly enhanced localized fields at the focal points. The size of the focal spots is less than half a wavelength and the focal positions are not determined by the Talbot distance.

In comparison to the results of above mentioned Talbot distances from scalar theory, only the contribution from the field of SPPs is taken into account in the calculations of the interference pattern. In the geometry considered in this paper, the conversion of the incident SPPs into the volume electromagnetic waves is weak. The Talbot distance of 2.016 µm (first order) calculated by formula $z_T=k\Lambda^2/\pi$ (it is the same as Eq. (13) actually. They were written in two different forms.) from Maradudin *et. al.* [63] and van Oosten [64] is different also from our calculated value of 0.3962 µm (first order) here. The reason we get a different answer compared to the other two methods is because the other two methods use the paraxial approximation in far-field regime, which state that the pitch of the structure is much greater than the incident wavelength. However, the near-field regime in which we operate is when the wavelength is about half the period. For our structures, the paraxial approximation does not hold anymore. Therefore, we can draw a conclusion that the theoretical equations deduced for the structures of metallic dot arrays [63, 65] are not suitable to our nanolens structures. The approach of computational numerical calculation must be employed for the study of the Talbot distance. For more information regarding theoretical calculation, readers can see Ref.[65].

To compare the experimental results with the theoretically calculated results, we plotted E-field intensity profiles versus propagation distance z together with that of the numerically calculated in the same figure, as shown in Fig. 37. It can be seen that variation tendency of the E-field intensity of the measured results is in agreement with the theoretically calculated results in general. As can be seen from Fig. 36, the positions of the measured first three focal points are slightly in advance of the calculated values due to the interaction of the metal surfaces between the NSOM probe and nanolenses at near-field. But for the last two ones,

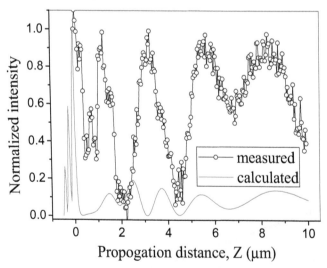

Figure 37. Comparison of E-field intensity distribution along propagation axis between experimental and theoretical results.

the measured and calculated results coincide with each other very well. It can be explained that the interaction between the NSOM probe and the nanolenses surface is weak at the far-field. The influence on the probing is significantly degraded and can be ignored accordingly.

In comparison to a standard diffraction experiment, if the wavelength is large the slit is subwavelength, surely there will be no diffracted waves (also known as cylindrical waves) and therefore, the distinction between short wavelength and long wavelength is obvious. For the standard diffraction, transmission intensity decays exponentially with slit size (Bethe's theory was put forth in 1944). However, for our plasmonic lenses, the transmission intensity can be enhanced due to inherent advantage of plasmonic resonance which is well known already. Here, we used metallic subwavelength structure to generate the self-imaging. It is more apparent in near field region. The contribution of plasmonic resonance here is that enhancing the intensity of the plasmonic Talbot-based focusing (see Fig. 36). It is especially important for nano-photolithography for which large working distance and high exposure intensity are crucial issues. Our previous FDTD calculation results [66] demonstrated that only one focal point can be observed for the case of $\lambda > 300$ nm, which is a threshold value for the SPPs-based Talbot effect. For the short wavelength, it is smaller than the slit width. Thus it is consistent to the Talbot effect condition. For more information regarding experimental characterization, readers can see Ref.[67].

7. Summary

In summary, the main reason of the diffraction limit of a conventional optical lens is that a conventional lens is only capable of transmitting the propagating. The evanescent carry subwavelength information about the object decay, but it decay exponentially and cannot be collected in far field region. In contrast, plasmonic lens is an alternative to these problems. There are two key concepts about plasmonic lens. One is the concept of negative refractive index. The other is the transmission enhancement of evanescent waves. The concept of negative refractive index is the basic issue of plasmonic lens. And this is very common for noble metal in specific frequency.

The scalar Talbot effect cannot interpret the Talbot effect phenomenon for the metallic nanolenses. It may attribute to the paraxial approximation applied in the Talbot effect theory in far-field region. However, the approximation does not hold in our nanolenses structures during the light propagation. In addition, the Talbot effect appears at the short wavelength regime only, especially in the ultra-violet wavelength region.

Also plasmonic lenses provide excellent optical property for us. We can shape the beam beyond the diffraction limit, such as superfocusing, imaging and so on. Also we can exist in various forms. And in this paper, we show them in three form generally. Some new ideas will be produced while we design the plasmonic lenses to resolve practical problems. Currently, applications of the plasmonic lenses reported include photodetector [67] and ring resonator [68]. With rapid development of the plasmonic lenses, it is reasonable to believe that extensive applications will be found in the near future.

Author details

Yongqi Fu

School of Physical Electronics, University of Electronic Science and Technology of China, China

Jun Wang and Daohua Zhang

School of Electronic and Electrical Engineering, Nanyang Technological University, Republic of Singapore

Acknowledgement

Some reported research works in this chapter were supported by National Natural Science Foundation of China (No. 11079014 and 61077010).

8. References

[1] J. B. Pendry, "Negative Refraction Makes a Perfect Lens", Phys. Rev. Lett. 5, 3966-3969 (2000).

[2] Anthony Grbic and George V. Eleftheriades, "Overcoming the diffraction Limit with a planar Left-Handed Transmission-Line Lens", Phys. Rev. Lett. 92, No. 11, 117403 (2004).

[3] D.R. Smith, J.B. Pendry, M.C.K. Wiltshire, "Metamaterials and negative refractive index", Science 305, 788-792 (2004).

[4] S. Anantha Ramakrishna, "Physics of negative refractive index materials", Rep. Prog. Phys. 68 (2), 449-521 (2005).

[5] T. W. Ebbesen, H. J. Lezec, H. F. Ghaemi, T. Thio and P. A Wolff, "Extraordinary optical transmission through subwavelength hole array", Nature 391, 667-669 (1998).

[6] H.J. Lezec, A. Degiron, E. Devaux, R.A. Linke, F. Martín-Moreno, L.J. García-Vidal, and T.W. Ebbesen, "Beaming light from a subwavelength aperture", Science 297, 820 (2002).

[7] Liang-Bin Yu, Ding-Zheng Lin et al., "Physical origin of directional beaming emitted from a subwavelength slit", Phys. Rev. B 71, 041405 (2005).

[8] W. L. Barnes, A. Dereux, and T. W. Ebbesen, "Surface plasmon subwavelength optics", Nature 424, 824-830 (2003).

[9] N. Fang, X. Zhang, "Imaging properties of a metamaterial superlens", Appl. Phys. Lett. 82(2), 161-163 (2003).

[10] N. Fang, H. Lee, C. Sun and X. Zhang, "Sub-Diffraction-Limited Optical Imaging with a Silver Superlens", Science 308, 534-537 (2005).

[11] Z. Liu, S. Durant, H. Lee, Y. Pikus, N. Fang, Y. Xiong, C. Sun, and X. Zhang, "Far-Field Optical Superlens", Nano Lett. Vol. 7, No. 2, 403-408 (2007).

[12] N. Fang, Z. W. Liu, T.J. Yen, X. Zhang, "Regenerating evanescent waves from a silver superlens", Opt. Express 11(7), 682-687 (2003).

[13] Z. W. Liu, J. M. Steele, W. Srituravanich, Y. Pikus, C. Sun, and X. Zhang, "Focusing surface plasmons with a plasmonic lens", Nano Lett. 5, 1726-1729 (2005).

[14] D.O.S. Melville, R.J. Blaike, C.R. Wolf, "Submicron imaging with a planar silver lens", Appl. Phys. Lett. 84(22), 4403-4405 (2007).

[15] Salandrino and N. Engheta, "Far-field Subdiffraction Optical Microscopy Using Metamaterial Crystals: Theory and Simulations", Phys. Rev. B 74, 075103 (2006).

[16] W. Nomura, M. Ohtsu, and T. Yatsui, "Nanodot coupler with a surface plasmon polariton condenser for optical far/near-field conversion", Appl. Phys. Lett. 86, 181108 (2005).

[17] Liang Feng, A. Kevin Tetz, Boris Slutsky, Vitaliy Lomakin, and Yeshaiahu Fainman, "Fourier plasmonics: diffractive focusing of in-plane surface plasmon polariton waves", Appl. Phys. Lett. 91, 081101 (2007).

[18] Z. Liu, H. Lee, Y. Xiong, C. Sun and X. Zhang, "Far-Field Optical Hyperlens Magnifying Sub-Diffraction-Limited Object", Science 315, 1686 (2007).

[19] A. V. Zayats, I. I. Smolyaninov, and A. A. Maradudin, "Nano-optics of surface plasmon polaritons", Phys. Rep. 408, 131-314 (2005).

[20] L. Zhou and C. T. Chan, "Relaxation mechanisms in three-dimensional metamaterial lens focusing", Opt. Lett. 30, 1812 (2005).

[21] H. Ditlbacher, J. R. Krenn, G. Schider, A. Leitner, F. R. Aussenegg, "Two-dimensional optics with surface plasmon polaritons", Appl. Phys. Lett. 81, 762-764 (2002).

[22] Yongqi Fu, W. Zhou, L.E.N. Lim, C. Du, H. Shi, C.T Wang and X. Luo, "Transmission and reflection navigated optical probe with depth-tuned serface corrugations", Appl. Phys. B 86, 155-158 (2007).

[23] H. F. Shi, C. T. Wang, C. L. Du, X. G. Luo, X. C. Dong, and H. T. Gao, "Beam manipulating by metallic nano-slits with variant widths", Opt. Express 13, 6815-6820 (2005).

[24] Avner Yanai and Uriel Levy, "The role of short and long range surface plasmons for plasmonic focusing applications", Opt. Express 17, No. 16 (2009).

[25] M. H. Wong, C. D. Sarris, and G. V. Eleftheriades, "Metallic Transmission Screen For Sub-Wavelength Focusing", Electronics Letters, 43, 1402-1404 (2007).

[26] Yongqi Fu, Wei Zhou, Lennie E.N. Lim, Chunlei Du, Haofei Shi, Changtao Wang, "Geometrical characterization issues of plasmonic nanostructures with depth-tuned grooves for beam shaping". Opt. Eng. 45, 108001 (2006).

[27] Baohua Jia, Haofei Shi, Jiafang Li, Yongqi Fu, Chunlei Du, and Min Gu, "Near-field visualization of focal depth modulation by step corrugated plasmonic slits", Appl. Phys. Lett. 94, 151912 (2009).

[28] Y. Xie, A.R. Zakharian, J.V. Moloney, and M. Mansuripur, Transmission of light through slit apertures in metallic films. Optics Express. 12. 6106-6121, (2004).

[29] L.B. Yu, D.Z. Lin, Y.C. Chen, Y.C. Chang, K.T. Huang, J.W. Liaw, J.T. Yeh, J.M. Liu, C.S. Yeh, and C.K. Lee, Physical origin of directional beaming emitted from a subwavelength slit. Physical Review B - Condensed Matter and Materials Physics. 71. 1-4, (2005).

[30] B. Ung and Y. Sheng, "Optical surface waves over metallo-dielectric nanostructures: Sommerfeld integrals revisited". Optics Express. 16. 9073-9086, (2008).

[31] H. Shi, C. Du, and X. Luo, "Focal length modulation based on a metallic slit surrounded with grooves in curved depths". Applied Physics Letters. 91, (2007).

[32] Z. Sun and H.K. Kim, "Refractive transmission of light and beam shaping with metallic nano-optic lenses". Applied Physics Letters. 85. 642-644, (2004).

[33] Yongqi Fu, W. Zhou, L.E.N Lim, C. Du, H. Shi, C. Wang and X. Luo, "Influence of V-shaped plasmonic nanostructures on beam propagation", Appl. Phys. B 86, 461-466(2007).

[34] Zhaowei Liu, Jennifer M. Steele, Hyesog Lee, and Xiang Zhang, "Tuning the focus of a plasmonic lens by the incident angle", Appl. Phy. Lett. 88, 171108 (2006).

[35] Jun Wang, Wei Zhou and Anand K. Asundi, "Effect of polarization on symmetry of focal spot of a plasmonic lens". Opt. Express 17, 8137-8143 (2009).

[36] LievenVerslegers, PeterB.Catrysse, ZongfuYu, JustinS.White, Edward S.Barnard, Mark L.Brongersma, and Shanhui Fan, "Planar Lenses Based on Nanoscale Slit Arrays in a Metallic Film". Nano Lett. 9, 235-238 (2009).

[37] Jennifer M. Steele , Zhaowei Liu , Yuan Wang, and Xiang Zhang, Resonant and non-resonant generation and focusing of surface plasmons using circular gratings. Opt. Express 14, 5664-5670 (2006).

[38] Yongqi Fu, W. Zhou, L.E.N.Lim, C.L. Du, X.G.Luo, Plasmonic microzone plate: Superfocusing at visible regime. Appl. Phys. Lett. 91, 061124 (2007).

[39] Yongqi Fu, Yu Liu, Xiuli Zhou, Zong Weixu, Fengzhou Fang, Experimental demonstration of focusing and lasing of plasmonic lens with chirped circular slits, Opt. Express 18 (4), 3438–3443 (2010).

[40] Wang, J., Zhou, W., Li, E. P. & Zhang, D. H., Subwavelength focusing using plasmonic wavelength-launched zone plate lenses. Plasmonics, 6, 269-272 (2011).

[41] Richards, B. & Wolf, E. Electromagnetic Diffraction in Optical Systems. II. Structure of the Image Field in an Aplanatic System. Proc. Roy. Soc. Lond A, Vol.253, No.1274, 358-379 (1959).

[42] Yongqi Fu, Wei Zhou, Hybrid Au-Ag subwavelength metallic structures with variant periods for superfocusing, J. Nanophotonics 3, 033504 (22 June 2009).

[43] Yongqi Fu, Wei Zhou, Lim Enk Ng Lennie, Nano-pinhole-based optical superlens, Research Letter in Physics, Vol.2008, 148505 (2008).

[44] Yongqi Fu, Xiuli Zhou, Yu Liu, Ultra-enhanced lasing effect of plasmonic lens structured with elliptical nano-pinholes distributed in variant period. Plasmonics, 5 (2), 111-116 (2010).

[45] Zhenkui Shi, Yongqi Fu, Xiuli Zhou, Shaoli Zhu, Polarization effect on focusing of a plasmonic lens structured with radialized and chirped elliptical nanopinholes. Plasmonnics 5(2), 175-182 (2010).

[46] Yu Liu, Yongqi Fu, Xiuli Zhou, Polarization dependent plasmonic lenses with variant periods for superfocusing. Plasmonics, 5(2), 117-123 (2010).

[47] Gilad M.Lerman, Avner Yanai,and Uriel Levy, "Demonstration of nanofocusing by use of plasmonic lens illuminated with radial polarized light". Nano Lett. 9, 2139-2143 (2009).

[48] Qiwen Zhan, "Cylindrical vector beams: from mathematical concepts to applications", Advances in Optics and Photonics 1, 1-57 (2009).

[49] T. Xu, C. L. Du, C. T. Wang, and X. G. Luo, "Subwavelength imaging by metallic slab lens with nanoslits", Appl. Phys. Lett. 91, 201501 (2007).

[50] A. Ono, J. Kato, S. Kawata, "Subwavelength optical imaging through a metallic nanorod array", Phys. Rev. Lett. 95, 267407 (2005).

[51] P. A. Belov, Y. Hao, S. Sudhakaran, "Subwavelength microwave imaging using an array of parallel conducting wires as a lens", Phys. Rev. B 73, 033108 (2006).

[52] Mark I. Stockman, "Nanofocusing of Optical Energy in Tapered Plasmonic Waveguides", Phys. Rev. Lett. 93, 137404 (2004).

[53] Ewold Verhagen, Albert Polman, and L. (Kobus) Kuipers, "Nanofocusing in laterally tapered plasmonic waveguides", Opt. Express 16, 45-57 (2008).

[54] A. B. Evlyukhin, S. I. Bozhevolny, A. L. Stepanov, R. Kiyan, C. Reinhardt, S. Passinger, and B. N. Chichkov, "Focusing and directing of surface plasmon polaritons by curved chains of nanoparticles", Opt. Exp. 15, 16667-16680 (2007).

[55] I. P. Radko, S. I. Bozhevolnyi, A. B. Evlyukhin, and A. Boltasseva, "Surface plasmon polariton beam focusing with parabolic nanoparticle chains", Opt. Express 15, 6576-6582 (2007).

[56] Leilei Yin, Vitali K. Vlasko-Vlasov, John Pearson, Jon M. Hiller, Jiong Hua, Ulrich Welp, Dennis E. Brown, and Clyde W. Kimball, "Subwavelength Focusing and Guiding of Surface Plasmons", Nano Lett. 2005, 5, 1399–1402.

[57] Z. J. Sun and H. K. Kim, "Refractive transmission of light and beam shaping with metallic nano-optic lenses", Appl. Phys. Lett. 85, 642 (2004).

[58] Z. J. Sun, "Beam splitting with a modified metallic nano-optics lens", Appl. Phys. Lett. 89, 26119 (2006).

[59] H. Ko, H. C. Kim, and M. Cheng, "Light transmission through a metallic/dielectric nano-optic lens", J. Vac. Sci. Technol. B 26, 62188-2191 (2008).

[60] S. Vedantam, H. Lee, J. Tang, J. Conway, M. Staffaroni, J. Lu and E. Yablonovitch, "Nanoscale Fabrication of a Plasmonic Dimple Lens for Nano-focusing of Light", Proceeding of SPIE 6641, 6641J (2007).

[61] James A. Shackleford, Richard Grote, Marc Currie, Jonathan E. Spanier, and Bahram Nabet, "Integrated plasmonic lens photodetector", Appl. Phys. Lett. 94, 083501 (2009).

[62] W. Srituravanich, L. Pan, Y. Wang, C. Sun, C. Bogy, and X. Zhang, "Flying plasmonic lens in the near field for high-speed nanolithography", Nature Nanotech. 3, 733-737 (2008).

[63] Dennis M R, Zheludev N I and García de Abajo F J , "The plasmon Talbot effect," Opt. Express, 15, 9692-700 (2007).

[64] D. van Oosten, M. Spasenovi, and L. Kuipers, "Nanohole Chains for Directional and Localized Surface Plasmon Excitation." Nano Lett. 10, 286-290 (2010).

[65] A A Maradudin and T A Leskova, "The Talbot effect for a surface plasmon polariton," New J. Phys. 11, 033004 (2009).

[66] Lingli Li, Yongqi Fu, Hongsheng Wu, Ligong Zheng, Hongxin Zhang, Zhenwu Lu, Qiang Sun, Weixing Yu, The Talbot effect of plasmonic nanolens, Opt. Express, 19(20), 19365-19373 (2011).

[67] Yiwei Zhang, Yongqi Fu, Yu Liu, Xiuli Zhou, Experimental study of metallic elliptical nanopinhole structure-based plasmonic lenses. Plasmonics, 6(2), 219-226 (2011).

[68] V. S. Volkov, S. I. Bozhevolnyi, E. Devaux, J.-Y. Laluet, and T. W. Ebbesen, "Wavelength selective nanophotonic components utilizing channel plasmon polaritons", Nano Lett. 7, 880 (2007).

Chapter 6

Novel SNOM Probes Based on Nanofocusing in Asymmetric Structures

Valeria Lotito, Christian Hafner, Urs Sennhauser and Gian-Luca Bona

Additional information is available at the end of the chapter

1. Introduction

Scanning near-field optical microscopy (SNOM) has become a microscopy tool of paramount importance for nanostructure investigation in different fields ranging from material to life science. In fact, such technology combines the advantages offered by classical optical microscopy with high resolution typical of scanning probe microscopy (SPM) [1,2]. More specifically, it allows overcoming the diffraction limit by exploitation of the non-propagating evanescent waves in the near-field zone, which are not retrieved in conventional optical microscopy. Still it retains some important features of this technique, e.g. the possibility to carry out real time observation of the sample in its native environment in a non-invasive way and with little sample preparation and, more importantly, the availability of a wide variety of contrast mechanisms, like polarization, absorption and fluorescence. Such advantageous characteristics are usually not affordable with non-optical scanning probe techniques, which require more awkward sample preparation and provide little information on non-morphological properties of the sample. On the other hand, similarly to SPM, SNOM enables to get also high resolution information on sample topography by scanning a sharp probe in the region of interest of the sample and using a non-optical short range interaction between the probe and the sample to control probe-sample distance. Hence, SNOM allows getting nano-scale insight into a specimen, casting light on both topographic and optical properties of the sample.

As suggested by the name itself, optical properties of SNOM rely on the near-field interaction between the probe and the sample: the probe can be used for near-field excitation of the sample, whose response is collected in the far-field or to detect the near-field response of the sample broadly illuminated in the far-field or for both. Independently of the specific excitation/detection scheme, a crucial role in the ultimate attainable optical resolution is played by the probe itself, whose design, modelling and optimization has been encouraged by a vast wealth of numerical methods, shedding light into the optical behaviour of the probe at very small scales.

The best known probe configuration, used by the pioneers of SNOM, is the so-called aperture probe, which is still commonly used today [1,3,4]. Such a structure is based on a metal-coated tapered dielectric (typically an optical fiber) with an aperture left at the very apex. In this case, resolution is mostly dictated by the size of the aperture, which cannot be decreased at will not only because of technological limits, but also due to the dramatic slump in signal throughput with decreasing aperture size. The latter cannot be simply improved by increasing the input power because of the potential risk for thermal damage. Significant heating can occur as a consequence of multiple back-reflections from the taper and metal absorption, resulting in aperture expansion, tip contraction and elongation, possible partial detachment or even breakdown of the metal layer due to mechanical stresses arising from the different thermal expansion coefficients of the fiber and the metal coating [1,5-7]. All these phenomena inherently cause degradation of the probe behaviour and impose severe limits on the injected power.

Another major factor influencing the performance of aperture probes is the taper cone angle. In fact, such probes can be viewed as a tapered hollow metal waveguide filled with a dielectric, in which the mode structure changes as a function of the characteristic dimension of the dielectric core [8]. Guided modes run into cut-off one after the other as the diameter decreases, until, at a well-defined diameter, even the last guided mode, the fundamental HE_{11} mode, runs into cut-off; for smaller diameters of the dielectric core, the energy in the core decays exponentially because of the purely imaginary propagation constant of all the modes [1]. The amount of light that reaches the probe aperture depends on the distance separating the aperture and the HE_{11} cut-off diameter: the larger the opening angle of the tapered structure, the better the light transmission of the probe is, as the final cut-off diameter approaches the probe apex [1].

Taper profile depends of course on the adopted fabrication method: for fiber-based probes, two main approaches are used for taper formation, either fiber heating and pulling (consisting in locally heating the fiber using a CO_2 laser and subsequently pulling apart the two parts) and chemical etching (which basically consists in dipping the fiber in a HF solution with an organic overlayer) [1,2]. The second approach usually results in shorter tapers, i.e. larger opening angles, beneficial for higher signal throughput in spite of a higher surface roughness, which can be reduced by resorting to some variants of the etching process like the so-called tube etching process. Even shorter tapers can be achieved by using the selective etching methods in which the different etching rates of the core and the cladding of the fiber are exploited to get probes in which only the core of the fiber is tapered, resulting in tapered length of few micrometers against the hundreds of micrometers of fibers produced by different techniques. The following steps consist in metal deposition (by thermal or e-beam evaporation or sputtering) and aperture formation, which can be done either by exploiting shadowing effects during metal evaporation or by punching or with focused ion beam (FIB) milling [1].

Several attempts have been carried out to overcome the fundamental limitations of aperture probes, i.e. poor throughput and poor resolution. Two main routes have been followed to improve their performance, i.e. optimization of the taper profile and of the aperture shape.

As pointed out earlier, chemical etching leads to an advantageous reduction of the overall taper length and a better control of the overall taper shape. Using selective chemical etching, probes based on multiple tapers like the double taper, the triple taper and the steeple-on-mesa taper profiles have been realized: the goal of these structures was to benefit from the advantages of large taper angles in signal throughput at least for part of the probe because the use of a single taper with a large taper angle would not be convenient during approach to the sample [9-13]. An asymmetric edged probe with a sharp edge at the foot of the conical taper has shown to improve the transmission of HE_{11} mode with proper linearly polarized excitation along the direction of the asymmetry [14]. Also the taper profile of probes produced by heating and pulling has been optimized: for this fabrication process the overall probe profile can be varied by controlling some process parameters; parabolic profiles have shown to give rise to higher throughput compared to conical profiles [15]. Theoretical studies have been carried out to assess the influence of taper profile on probe performance to determine optimal shapes in terms of throughput: for example, structures based on the alternation of conical and cylindrical sections or analytical expressions for improved taper profiles have been proposed as well as corrugations at the interface between the dielectric core and the metal coating or on the external metal coating; in the latter two cases, the increase in the throughput is due to the exploitation of surface plasmon polaritons (SPPs) [16-19].

Another route followed for the improvement of the performance of aperture probes has been the implementation of other aperture shapes different from the typical circular design: rectangular, square, slit, elliptical, C-shaped, I-shaped or dumbbell, H-shaped, bowtie, connected and separated double aperture, triangular, rod hole and tooth hole, gap apertures have been analysed and/or fabricated in fiber- and cantilever-based probes [20-28]. The improved throughput and field localization for some preferential input polarizations is due to the strong asymmetry of such aperture shapes and to the excitation of SPPs.

Despite the efforts in the optimization of the aperture probe to overcome its fundamental limits in throughput and resolution, the current design of aperture probes is not able to sustain routinely ultrahigh resolution imaging because of the limit of the skin depth of the metal coating that increases the effective aperture size, thus making the field distribution at the aperture significantly larger than its physical size [29]. Better throughput and resolution are expected if an apertureless metal or fully metal-coated probe is used, due to the combination of electrostatic lightning rod effect and surface plasmon excitation for proper illumination conditions. For example, due to the lightning rod effect, strong enhancement is expected at the tip apex, even though a real probe apex does not represent a real singularity (i.e. not defined first and second order derivatives) because of the finite conductivity of real metals and the finite tip radius [29]. In the next paragraph, we will see how plasmonic effects in apertureless probes can be used to get high resolution.

2. Nanofocusing in SNOM probes: The axisymmetric structure

Apertureless probes represent a promising alternative probe configuration. The fundamental problem with such probes is given by the fact that they are usually illuminated externally either by a focusing lens or using a prism-based total internal reflection configuration, resulting in a strong background due to far-field illumination, which could be detrimental for some sensitive samples and can be reduced only partially by on-axis illumination [30]. In order to reduce such background some groups have envisaged different solutions like far-field excitation of SPPs at the wide end of a tapered metal nanowire by using grating geometry: such SPPs propagate towards the apex of the nanowire leading to nanofocusing, that is delivery and concentration/focusing of the optical energy at the nanoscale, which means a region much smaller than the dimensions allowed by the diffraction limits [30,31].

Nanofocusing in metal tapered nanorods has been theoretically studied starting from the analysis of the SPP modes of a plasmonic waveguide consisting of a metal nanowire whose axis coincides with the z axis and whose dielectric function $\varepsilon_m(\omega)$ is uniform in space with ω being the excitation angular frequency; the wire is surrounded by a dielectric with dielectric function ε_d [31]. The three fundamental SPP modes supported by a metal nanowire are the transverse magnetic TM mode and two hybrid modes HE1 and HE2 (Figure 1). Note that we adopted this notation to indicate the modes, although the classification in TM0n (or TMn), TE0n (or TEn), HEmn, EHmn (where m is related to the angular symmetry and n to the radial variation of the mode) commonly used for circular dielectric wires might still be extended to circular metallic wires. Moreover, we considered the modes with lower losses.

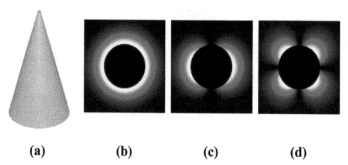

(a)   **(b)**   **(c)**   **(d)**

Figure 1. Sketch of: (a) fully metal probe; normalized electric field distributions of (b) TM plasmon mode, (c) HE1 plasmon mode, (d) HE2 plasmon mode.

The TM SPP mode can exist for arbitrarily small diameters of the nanowire and, hence, has been shown to be suitable for nanofocusing. This mode is axially uniform and is characterized by a magnetic field having only ϕ component; both electric and magnetic fields are independent of ϕ. Assuming that the radius $R(z)$ of the nanowire is a smooth function of z (which decreases from microscale to nanoscale for z going from large negative values to zero), the so-called eikonal or Wentzel-Kramers-Brillouin (WKB) or quasi-classical or geometric optics approximation can be applied to study the SPP modes and back-reflections can be neglected [31]. In order to study their propagation in a tapered metal rod, one can use a staircase approximation and interpret the taper as a series of cylindrical nanowires with smoothly decreasing diameters. The effective index $n_{eff}(R) = n_{eff}(R(z))$ for the TM mode of the plasmonic waveguide at a point z can be determined from:

$$\left[\frac{\varepsilon_d}{\kappa_d}\frac{K_1(\kappa_d R)}{K_0(\kappa_d R)} + \frac{\varepsilon_m}{\kappa_m}\frac{I_1(\kappa_m R)}{I_0(\kappa_m R)}\right] = 0 \tag{1}$$

where I_n and K_n are the modified Bessel functions of the first and second kind, $\kappa_d = k_0\sqrt{n_{eff}^2 - \varepsilon_d}$ and $\kappa_m = k_0\sqrt{n_{eff}^2 - \varepsilon_m}$. Under the plasmonic condition $\mathrm{Re}\{\varepsilon_m\} < -\varepsilon_d$, one gets the solutions for the SPP modes. For a thick wire ($k_0 R \gg 1$), the solution is the same as for a flat surface:

$$n_{eff} = \sqrt{\frac{\varepsilon_m \varepsilon_d}{\varepsilon_m + \varepsilon_d}} \tag{2}$$

For a thin, nano-scale radius wire ($k_0 R \ll 1$) one gets approximately with logarithmic precision [31]:

$$n_{eff}(R) \approx \frac{1}{k_0 R}\sqrt{-\frac{2\varepsilon_d}{\varepsilon_m}}\left[\ln\sqrt{-\frac{4\varepsilon_m}{\varepsilon_d}} - \gamma\right]^{-1} \tag{3}$$

where $\gamma \approx 0.57721$ is the Euler constant. From the previous expression it can be inferred that at the tip ($k_0 R \to 0$), $n_{eff} \to \infty$; therefore, the wave number increases and the SPPs asymptotically stop as both the phase and group velocity tend to zero because they are both proportional to $k_0 R$ [31]. Besides, the study of the propagation of the SPPs through the tapered metal rod carried out with the staircase approximation has revealed an anomalous increase in the SPP field amplitudes, as SPPs approach the tip, because the electric field varies as $\propto R^{-\frac{3}{2}}$ [32]. As a consequence, the simultaneous wavelength decrease and amplitude increase lead to a concurrent energy localization and, hence, to nanofocusing. Several theoretical studies have shown such a phenomenon [32-35]. The TM SPP mode can be excited using an axially symmetric grating under radially polarized excitation [36,37] or double-sided E-symmetric excitation [38] or asymmetric excitation via grating coupling on just one side of the tip [39].

An even more promising solution compared to fully metal probes in terms of simplification of the experimental set-up is the excitation of SPP modes on the metal coating of a fully metal-coated dielectric probe used under internal back excitation [29,40].

As for fully metal probes, the adiabatic approximation has been used to evaluate analytically the wave propagation in the fully metal-coated tapered dielectric probe, considering the cone as a succession of cylinders made up of a dielectric core of radius r_1 surrounded by a metal coating of thickness δ located in its turn in a dielectric medium [41]: in a first approximation, waveguide modes (WGM) have been calculated as those supported by a dielectric core surrounded by an infinitely thick metallic coating neglecting the external dielectric, while SPP modes have been calculated as those supported by a metal wire surrounded by a dielectric neglecting the internal dielectric core. Dispersion relations for waveguide modes and for SPP modes have thus been determined (the latter coincides with the dispersion equation for the metal wire). At a certain value of r_1, the wavevector of the WGM can match the one of the SPP mode: in these conditions, energy transfer from the WGM into the SPP mode is possible and, as a result, surface plasmons are excited bringing about nanofocusing. The transfer of energy and the field profiles can be determined with this approach using the coupled mode theory [41,42]. In this way, the SPPs can be excited using the WGM, thereby overcoming the problem of background inherent in the use of fully metal probes under external illumination.

The energy transfer from WGM to SPP modes has been thoroughly investigated in several studies, taking into account not only the "outer" SPP modes at the outer metal surface, but also the "inner" SPP modes at the inner metal surface; in some studies, non-conical taper profiles have been scrutinized as well [40,43-45]. The WGM has to present proper characteristics in order for the excited surface plasmons to exhibit axial symmetry, thereby interfering constructively at the tip apex and therefore generating nanofocusing. The three fundamental WGMs are a pair of orthogonal linearly polarized modes followed by a radially polarized one. If the fully metal-coated structure is excited by a linearly polarized HE11 mode, then the electric fields of the excited surface plasmons have opposite polarities on the opposite sides of the probe, giving rise to destructive interference at the very end of the probe; on the contrary, a radially polarized WGM excites SPPs with axial symmetry, which interfere constructively at the tip apex. These mechanisms are responsible for the high field confinement in case of radially polarized excitation compared to the linearly polarized case. The eventual outcome of this process is the creation of an ultrasmall hot spot in the region close to the tip apex in the former situation, as opposed to broader and weaker two-lobed electric field intensity distributions for the latter one. In particular, the size of the achievable hot spot in case of radially polarized excitation (and, hence, the ultimate attainable resolution) is mostly limited by the diameter of the metal apex, which can be decreased at will. Such behaviour has been confirmed in both theoretical and experimental studies [29,40,46-50] and is sketched in Figure 2.

Because of its attractive characteristics, the fully metal-coated dielectric probe has been chosen as the starting point for structural optimization; as explained in more detail in the next paragraph, the goal of the work will be the achievement of field localization (a feature of utmost importance for SNOM applications) under linearly polarized excitation, bringing about a further substantial simplification in experimental set-ups.

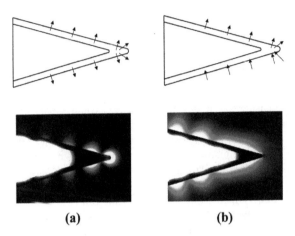

(a)   **(b)**

Figure 2. Sketch of the behaviour of a fully metal-coated dielectric probe under: (a) radially; (b) linearly polarized excitation. Arrows indicate the electric fields associated with the surface plasmons (adapted from [46]).

3. Nanofocusing in SNOM probes: The asymmetric structure

As previously explained, axisymmetric fully metal-coated probes under internal back excitation hold the promise for high-resolution applications, as they can allow the achievement of a strongly localized hot spot, whose size is mostly limited by the diameter of the metal apex, which can be decreased at will. However, as pointed out, the resolution of fully metal-coated tips is highly sensitive to the polarization state of the input field: such desirable field localization properties are affordable only under radially polarized excitation.

The reason for the strong polarization-dependent behaviour is due to the different characteristics of the surface plasmons excited by the different waveguide modes on the external metal coating: those excited by a linearly polarized mode interfere destructively at the tip apex, resulting in a weak and broad near-field distribution contrarily to those excited by a radially polarized mode which interfere constructively bringing about field localization.

The fundamental drawback of radially polarized excitation, despite its potential attractiveness, resides in its cumbersome injection procedure, which is extremely sensitive to misalignments that could impair the potential benefits stemming from its use [50,51].

In order to get similar superfocusing effects under a more easily excitable linearly polarized mode, one could break the axial symmetry of the fully metal-coated probe so as to avoid the destructive interference between the SPPs excited by a linearly polarized mode on the opposite sides of the axisymmetric structure. If z is the direction of the probe axis and an asymmetry is introduced in the tip structure along x, field confinement under x linearly polarized mode could be expected. This idea stems from the numerical investigation of the effects of unintentional asymmetries like single and multiple air spherical bubbles, which have been shown to have a weak field localization effect due to the coupling between the linearly polarized mode along the direction of the asymmetry and the radially polarized one [52].

Of course, superfocusing based on random defects could not be easily forecast, but stronger and more easily predictable focusing effects can be achieved by the introduction of intentional modifications. The asymmetry can be either present in the probe structure itself or in the probe illumination scheme. Unilateral and bilateral slits in the metal coating have been numerically studied as a form of structural modification. Tip on aperture probes and probes based on a monopole antenna can also be included in this category [53,54]. An asymmetric illumination scheme can be used as an alternative to a structural asymmetry [55]: surface plasmons are excited only on one side of an apertureless probe via an opening close to the probe base in the offset aperture probe [56], while asymmetric single-sided SPP excitation results in field localization in an axisymmetric apertureless probe [38].

We have to underline how all these asymmetric structures are based on the introduction of an asymmetry along one specific direction and, for this reason, we can indicate them as "directional" asymmetries [55,57]. Hence, field localization occurs only for the linearly polarized mode that is oriented in the direction of the asymmetry: for example, if the asymmetry in the structure or in the illumination is along x, field localization is expected under x linearly polarized excitation, but not under y linearly polarized excitation, because no asymmetry is present along the y direction. This is due to the fact that all these forms of asymmetries exhibit a plane of symmetry and, hence, cannot guarantee field localization for arbitrarily oriented linearly polarized modes. Such asymmetries still require alignment between the linearly polarized mode and the asymmetry itself, even though injection procedures are considerably simplified because radially polarized excitation is no longer necessary.

The practical implication of such a behaviour in experimental applications is apparent if considering the difference in imaging properties using asymmetric tips based on a tip-on-aperture and on a monopole antenna grown on the rim of an aperture tip with different orientation of the input linear polarization, with dramatic variations in resolution and signal intensity upon rotation of the input excitation from the direction of the asymmetry to the orthogonal one [53,54]. However, the control of the direction of the input polarization is not an easy task and often imposes a determination *a posteriori* of the effective polarization direction close to the asymmetry by comparison of the field distribution observed close to the tip apex with simulation results [14].

Therefore, it would be desirable to attain superfocusing effects for arbitrarily oriented linearly polarized excitation, by using an "adirectional" asymmetry, which means a suitable modification likely to create an asymmetry along all spatial directions [55,57].

In the coming paragraphs, we will report the numerical analysis of novel forms of asymmetry based both on new types of structural directional asymmetries and on the pioneering concept of adirectional design. In particular, we will illustrate the properties of:

1. directional asymmetries based on: (i) an oblique cut; (ii) asymmetric corrugations in the metal coating;
2. adirectional asymmetries based on: (i) a spiral corrugation; (ii) azimuthal corrugations arranged in a spiral-like fashion.

Before tackling the design and the characteristics of these specific structures, in the next paragraph a short overview of the issues and difficulties encountered in probe modelling is presented.

4. Optimization of probe structures: Challenges of tip modelling

The theoretical study of the behaviour of SNOM probes is essential not only to get insight into the characteristics of commonly used probe structures and identify potential problems in imaging, but also to detect possible routes for optimization and predict the implications of the use of novel probe configurations. With such an approach, experimental efforts can be devoted to those probes that theoretically exhibit the most promising features.

Analytical solutions have been determined by Bethe and Bouwkamp for an aperture in an infinitely thin perfectly conducting screen, which laid down the foundation for further theoretical treatments. In the search for a model that could describe more faithfully the aperture probe behaviour, analytical studies that took into account the finite thickness of the metal screen have been carried out [58-59]. All of these treatments, however, suffer from neglecting the finite conductivity of the metal cladding used in real SNOM probes and provide little resemblance to the actual tip geometries used in experiments [60].

Other semi-analytical approaches have been based on a staircase approximation in which the longitudinally non-uniform waveguide (the tapered part of the probe) is considered as a succession of cylindrical sections of decreasing radius and the eigenmodes of the uniform waveguides obtained by infinitely stretching along the axis of the probe at each cross section are computed, as illustrated earlier. An analytical evaluation of the power transmitted by an aperture probe based on such an approximation using a mode matching theory has been done [1]. Starting also from this staircase approximation, different probe profiles like parabolic, exponential and mixed shapes based on the alternation of conical and cylindrical sections along the taper have been examined using the cross-section method [16, 17].

As discussed in paragraph 2, the staircase approximation has also been used for the analysis of nanofocusing in fully metal and fully metal-coated dielectric probes using local mode theory in a weakly non-uniform optical waveguide and the eikonal approximation [31, 34, 40, 43, 44].

Although analytical expressions are useful to get an understanding of probe behaviour, their range of applicability is limited. For example, the eikonal approximation can be applied as long as the adiabatic criterion is satisfied, i.e. for small local taper angles

[31,34,40,43]. On the other hand, the study of probes that do not satisfy such requirement can be of great interest: for example, optimal conditions for nanofocusing on tapered metal rods have been found in nonadiabatic conditions, which fail to be treated analytically [61]. Moreover, analytical methods turn out to be not adequate to model abrupt variations in the tapered profile as they often suppose a weak longitudinal non-uniformity.

Whenever more challenging probe geometries need to be studied, a numerical approach becomes mandatory. For this reason, the development of optimized probe structures has been accompanied and favoured by the flourishing of a vast range of numerical methods giving a glimpse at nanoscale mechanisms [9,62-64].

Such methods have been extensively used to investigate novel aperture shapes using, for example, the finite difference time domain method (FDTD) or the field susceptibility technique and in most of the cases considering apertures in thin metal films [21,24,26-28]. Furthermore, the analysis of wave propagation in tapered structures (either traditional conical tapers or structures modified with corrugations or multiple tapers) with an aperture at the end has been carried out using the finite difference beam propagation method [65], the FDTD method [18,19], the body of revolution FDTD method (BOR-FDTD) [48], the multiple multipole method (MMP) [66]. Fully metal probes and fully metal-coated pyramidal probes with different shapes have also been analysed both in case of direct illumination at the metal apex or with far-field excitation further away from the metal apex as described in paragraph 2 [36,37,61,67] using FDTD, BOR-FDTD, finite integration technique (FIT), and the finite element method (FEM). The fully metal-coated dielectric structure which, as anticipated, is the one of interest in this chapter, has been intensively numerically investigated especially under internal back excitation and its polarization-dependent properties have been carefully examined: the need for a radially polarized excitation has been pointed out as essential to get field localization [40,46-50].

Even numerical treatment is challenging because near-field probes involve different length scales: while phenomena of major interest for near-field interactions occur in the mesoscopic (sizes of the order of the incident wavelength) and nanoscopic (structures smaller than 100 nm) regimes [63], the overall probe structure, especially for fiber-based probes, can include sections much bigger than the incident wavelength. Computational cost for modelling a probe in its entirety would become prohibitive. Therefore, the overall computational domain is typically restricted to the very end of the probe, often using two-dimensional approximations or exploiting symmetry properties of the structure.

In the light of the previous considerations, we have investigated our novel probes based on appropriate modifications introduced in fully metal-coated dielectric probes adopting a numerical approach and restricting the analysis to the very end of the probe. An overview of the numerical approach adopted for our simulations is provided in the next paragraph, together with the description of the developed computational model. The model has been tested first to study the behaviour of the fully metal-coated probe, which, as reported in paragraph 2, has been already numerically and experimentally investigated and represents the reference for comparison for the new probe configurations.

5. Description of the computational model

Among the different numerical methods used in the past to simulate fully metal-coated SNOM probes, ranging from the MMP [46,68], to FIT [40,49,50,52,69], or the FDTD method [48], we have preferred FEM, chosen also in [47].

FEM is a tool used for the solution of differential equations in many disciplines, ranging from electromagnetics to solid and structural mechanics, from fluid dynamics to acoustics and thermal conduction [70]. A point of strength of FEM is its ability to deal with a complex geometry. Unstructured grids can accommodate for complex geometries in a much more straightforward way than other methods using Cartesian grids like finite difference methods.

The irregular domain is discretized into smaller and regular subdomains, known as finite elements, thereby replacing a domain having an infinite number of degrees of freedom by a system with a finite number of degrees of freedom. The essential principle behind FEM is a piecewise approximation: the solution of a complex problem is obtained by splitting the region of interest into smaller regions and approximating the solution over each subregion by a simple function [71].

In particular, our three-dimensional (3D) computational model for the simulation of the electromagnetic modes in the investigated probe configurations has been developed with the help of a commercial software (Comsol Multiphysics) based on FEM. The computational process is articulated in a two-dimensional (2D) analysis to calculate the eigenmodes at the input port, followed by the 3D simulation of the propagation of the first three eigenmodes, i.e., the two lowest order linearly polarized modes and the radially polarized one. Second order elements with minimum size of about 0.8 nm have been used. Simulations have been run on a 64 bit workstation with 32 GB of RAM.

The first examined structure is the fully metal-coated dielectric probe under radially and linearly polarized excitation. Although in this case one could benefit from the symmetry properties of the structure, the axisymmetric probe represents only the starting point for the search of optimized configurations, illustrated in the following paragraphs. As we will see, such structures are characterized by strong asymmetries, which impose the need for full 3D analysis. Therefore, for the sake of a better comparison, a 3D modelling has been adopted also for the reference axisymmetric structure, without resorting to any of the simplifications used in previous works to handle axially symmetric structures. Only the very end of the tip is examined due to the high computational burden of the simulations because of the different scales of the metal layer and the dielectrics. In previous works, simulations involving larger portions could be carried out only when the structure was less computationally challenging due the rotational symmetry of the probe allowing reduction of the problem complexity either with the BOR-FDTD method or even with approximate 2D simulations [18,48].

Figure 3 reports the sketch of the simulated axisymmetric probe, consisting of a silica core (n=1.5) surrounded by an aluminium coating (n=0.645+5.029i at the operating wavelength

$\lambda=532$ nm). The radii of the inner silica cone and of the outer metallic hollow cone are 225 nm and 275 nm, respectively. Both cones (having an apex angle of 30°) are rounded, with the radius of curvature of the inner cone being 10 nm, the one of the outer cone amounting to 20 nm. The overall modelling domain is a 1.6 µm cylinder with radius 1 µm. The probe axis lies on the z axis.

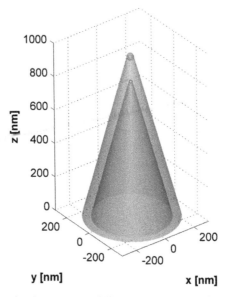

Figure 3. Sketch of the simulated axisymmetric fully metal-coated dielectric probe.

In Figure 4, the first three eigenmodes, i.e. the two lowest order x and y linearly polarized modes (H and V) and the radially polarized one (R), are reported together with the corresponding near-field distributions (square of the norm of the electric field) taken over a square area 600 nm by 600 nm centred around the probe apex at 10 nm from the apex; normalization to the maximum value of the electric field intensity distribution for the radial polarization has been done in order to emphasize the relative field strengths [51].

A highly localized hot spot with a full width at half maximum (FWHM) of 38 nm is observed for the R mode due to the constructive interference of the surface waves along the taper. As shown in [49], the size of the hot spot is influenced by the final rounding in the metal coating that, in our simulations, was chosen to be 20 nm in radius just for convenience as a reference for comparison and can be decreased at will. On the contrary, destructive interference of surface waves at the tip apex gives rise to two-lobed distributions, polarized mainly along the x and y axis, under H and V linearly polarized excitation, respectively; the average size (measured as the distance over which the field is more than or equal to half of its peak value) is approximately 400 nm and the peak value is about 50 times smaller than the peak of the R spot.

These results, in agreement with those previously obtained for such a probe, represent the reference for comparison for the novel probe configurations scrutinized in the following paragraphs, in which the characteristics of the near-field distributions at 10 nm from the apex of the modified probes are normalized to the peak value obtained in the axisymmetric probe under R excitation for an easier comparison.

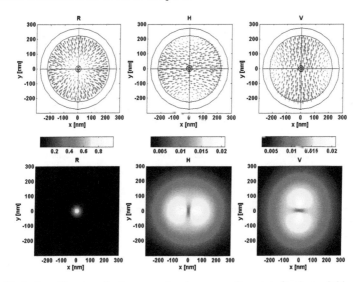

Figure 4. Illustration of input modes (upper row) and corresponding normalized near-field distributions at 10 nm from the apex of a standard axisymmetric fully metal-coated probe (lower row) (adapted from [51]).

6. Numerical investigation of probes with directional asymmetries under linearly polarized excitation

As anticipated, we have designed and investigated the behaviour of two different structural modifications that can be classified as directional asymmetries, one based on an oblique cut close to the tip apex stripping off both the metal coating and the inner core, and the other consisting of asymmetric corrugations in the metal coating. The sketches of the two structures are reported in Figure 5.

The first tip (Figure 5(a)) is cut along a plane which is neither orthogonal nor parallel to the tip axis: the cut angle (defined as the angle between a plane orthogonal to the axis of the tip and the plane of the cut itself) and the cut height (meant as the height of the new tip apex after the cut, measured from the bottom of the computational domain) can be varied [72].

The structure with asymmetric corrugations (Figure 5(b)) is based on the introduction on the outer metal surface of semicircular corrugations (either bumps or grooves), modelled by joining five truncated toroids of radius 20 nm with hemispherical terminations having the same radius: the toroidal sections are filled with air in case of a groove or with metal in case of a bump [51]. The case of metal oxide filling has been considered as well. Corrugations are limited to just one half of the tip, hence their angular extension is less than 180°.

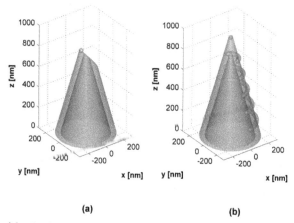

(a)   (b)

Figure 5. Sketch of the simulated structures based on an adirectional asymmetry: (a) probe with oblique cut; (b) asymmetrically corrugated fully metal-coated probe.

In both the cases, the asymmetry is present along the x direction.

6.1. Probe with oblique cut

For this probe structure, the behaviour of the probe under variable cut angle at constant cut height and for variable cut height at constant cut angle has been analysed [72].

In the first case, the angle has been varied from 20° to 60° (with a step of 10°) at a height of 816 nm. Normalized near-field patterns under H and V polarized excitation are shown in Figure 6.

As expected, field localization is achieved only under H polarized excitation. The two initially separated lobes of the V mode tend to merge as the angle becomes steeper until getting completely intermingled, while the initially asymmetric spots obtained under H excitation become progressively more symmetric about their centre along the x axis as the cut angle increases. The FWHM and the peak value normalized to the one obtained in an axisymmetric probe under radially polarized excitation are reported in Figure 7 as a function of the cut angle. No dramatic changes occur in the size of the spots (always between 37 nm and 41 nm), while an increase in the H peak occurs with steeper cut angles.

A similar analysis carried out for a cut height varying from 741 nm to 841 nm (with a step of 25 nm) at a cut angle of 30° revealed once again field localization under H polarized excitation. The results of the simulations with variable cut height suggest the use of cuts involving a larger fraction of the originally axisymmetric probe because both the FWHM and the peak value of the achieved hot spot undergo deterioration with increasing cut height. However, the approach to samples with steeper topographic variations would be hampered by cuts at lower heights unless a large cut angle is used at the same time, imposing a trade-off with the quality of the achievable H hot spot in the choice of the cut height.

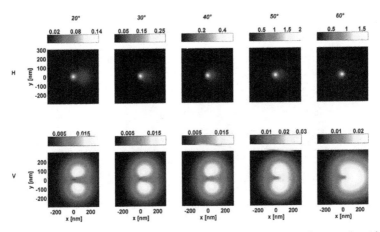

Figure 6. Normalized near-field intensity distributions at 10 nm from the apex of a cut probe with cut height of 816 nm and cut angle variable from 20° to 60° under H (upper row) and V (lower row) excitation (adapted from [72]).

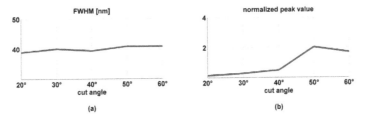

Figure 7. Characteristics of the near-field intensity distributions at 10 nm from the apex of the cut probe under H polarized excitation for variable cut angle: (a) FWHM; (b) peak value normalized to the one achieved in an axisymmetric probe under radially polarized excitation (adapted from [72]).

6.2. Probe with asymmetric corrugations

As we said, the five corrugations could consist in either grooves or bumps on the outer metal coating and are equally spaced, with the z-spacing amounting to 150 nm and the first bottom one centred at 150 nm from the input port [51].

First, the effect of a variation in the sequence of grooves and bumps has been considered and all the possible permutations of air indentations and metal bumps for the same structure have been analysed (at constant azimuthal extension of 160°). The different configurations have been named after the initial of the filling material (*a* for air groove and *m* for metal bump) starting from the bottom corrugation. As an example, the normalized near-field distributions for the configuration *amama* is reported in Figure 8. As expected, field localization is achieved under H excitation, while the V distribution maintains an almost two-lobed pattern.

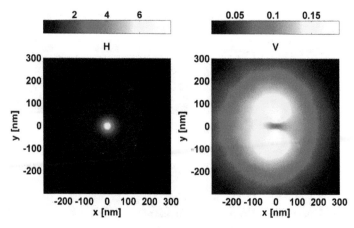

Figure 8. Normalized near-field intensity distributions at 10 nm from the apex of a probe with asymmetric corrugations (*amama* configuration) (adapted from [51]).

A systematic analysis of the FWHM and the peak value of the hot spot achieved under H polarized excitation for all the possible permutations of air indentations and metal bumps for the same structure has revealed that, except for only one configuration, the size of the near-field distribution undergoes a significant shrinkage in all the cases with the creation of a real ultrasmall spot in most of the cases with a FWHM comparable to that observed for the R mode excitation in the axisymmetric probe (Figure 9(a)). Although the H peak value generally increases compared to the case of the standard axisymmetric probe, only few material combinations give rise to values comparable or, in two cases, even much superior to the radial peak for the axisymmetric probe, with the best results given by the *amama* configuration (Figure 9(b)). Similar trends in terms of alternation of metal and dielectric have been observed upon replacement of the air grooves with metal oxide bumps (in this case aluminium oxide) and considering all the possible permutations of metal and metal oxide in the five semirings. However, the substitution of air with metal oxide gives rise to higher peak values due to better coupling between inner and outer SPPs when air is substituted with metal oxide, because the coupling of surface modes at two adjacent metal-dielectric interfaces becomes more efficient when the indices of refraction of the two dielectrics are closer [73].

Next, the influence of a variation of a geometric parameter, i.e. the azimuthal extension of the corrugations, has been studied. The *amama* configuration has been chosen and the angular extension of each corrugation has been changed between 110° to 160° with a step of 10°. No significant variations in the FWHM for the H hot spots have been observed, while the peak value increases with increasing azimuthal extension of the corrugations.

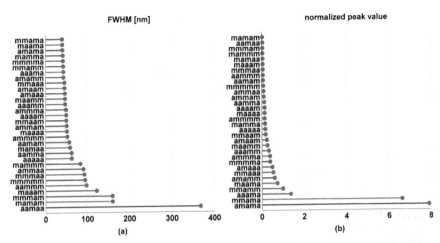

Figure 9. Characteristics of the near-field intensity distributions at 10 nm from the apex of the probe with asymmetric corrugations under H polarized excitation for each of the material permutations in the five semirings: (a) FWHM; (b) peak value normalized to the one achieved in an axisymmetric probe under radially polarized excitation (adapted from [51]).

6.3. Analysis of the behaviour of the directional asymmetries for arbitrary orientation of the asymmetry with respect to the input polarization

Our finite element based simulations have shown that carefully designed asymmetries introduced in an originally axisymmetric fully metal-coated tip can produce field localization under an excitation linearly polarized along a proper polarization direction. The presence of the asymmetry causes the electric fields associated with SPPs on the opposite sides of the tip not to have opposite phases any longer, a phenomenon that leads to destructive interference under linearly polarized injection in an axisymmetric structure [51,57]. A global analysis of the results obtained for both a cut probe and a tip with asymmetric corrugations has shown that this effect is enhanced when the asymmetry is extended over a broader region (as is the case for steeper cut angles or lower cut heights in case of the probe with oblique cut or a larger azimuthal extension of the corrugation for the tip with asymmetric corrugations). In this way, superfocusing can be achieved with a linearly polarized injection, which is much easier than a radially polarized one, with an enormous simplification in experimental applications.

However, so far we have considered the behaviour of the probe for two specific orientations of the input linearly polarized excitation with respect to the asymmetry (located along x), i.e. alignment along the direction of the asymmetry (H polarization) or alignment along the direction orthogonal to the asymmetry (V polarization). Field localization occurs for input polarization aligned along the preferential direction of the asymmetry, while no significant variation compared to the axisymmetric probe is shown for input polarization orthogonal to such direction.

In this section, to better understand how the misalignment from the preferential spatial direction can affect the performance of a probe with a directional asymmetry, we show the behaviour of a probe based on asymmetric corrugations for variable mutual orientation of the direction of the input linearly polarized excitation with respect to the one of the asymmetry [55,57,74]. Such mutual alignment has been defined as the angle α between the direction of the input linear polarization and the angle bisector of the corrugations (Figure 10(a)); the mutual orientation specified by α was varied from -85° to 90° to encompass all the possible mutual positions. Note that $\alpha=0°$ and $\alpha=90°$ correspond to x and y linearly polarized excitations so far labelled as H and V, with $\alpha=0°$ representing the position of best alignment of the input linear polarization with respect to the directional asymmetry and $\alpha=90°$ representing the maximum misalignment.

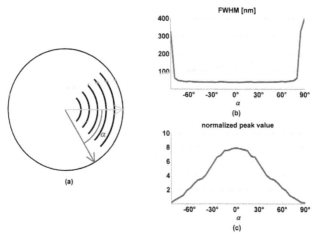

Figure 10. Behaviour of the probe with asymmetric corrugations for variable orientation of the input linearly polarized excitation with respect to the asymmetry: (a) schematic of the xy projection of the probe: the angle α is the one between the direction of the input linear polarization (magenta line) and the angle bisector of the corrugations (cyan line); (b) FWHM of the near-field distribution for the *amama* configuration under variable α; (c) peak value of the near-field distribution normalized to the one of the axisymmetric probe under radially polarized excitation for the *amama* configuration under variable α (adapted from [55]).

Figure 10 (b) and (c) report the FWHM and the normalized peak value as a function of α for the *amama* configuration and azimuthal extension of the corrugations equal to 160°. As the misalignment of the input polarization from the preferential direction of the asymmetry increases, both the peak value and the shape of the near-field intensity pattern change. In particular, the peak value decreases and the distribution becomes gradually broader. This is due to the fact that, as α increases, the asymmetry perceived by the input linearly polarized excitation progressively disappears. In retrospect, the different behaviour can be explained if recalling that a structure asymmetric along x appears symmetric for a y linearly polarized excitation, which brings about destructive interference of the excited SPPs, similarly to what happens for an axisymmetric structure under linearly polarized excitation. If we consider the average of both the FWHM and the peak value over all the possible mutual positions, from the graphs it can be inferred that maximum deviations of the peak value and of the spot size from the corresponding averages are about 100% and 450%, respectively.

Similar analyses have been run for various forms of directional asymmetries with different structural parameters. The maximum deviation of the peak ratio and of the spot size from the average value over all the mutual orientations of the linearly polarized excitation with respect to the asymmetry is reported in order to highlight the sensitivity to mode orientation (Figure 11).

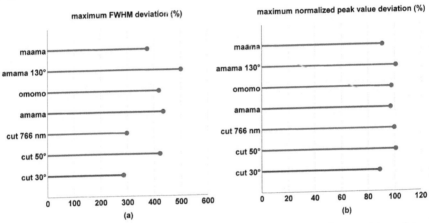

Figure 11. Characteristics of the near-field intensity distributions for directional asymmetries under linearly polarized excitation for variable orientation α of the asymmetry with respect to the input linear polarization: (a) maximum deviation from the average FWHM; (b) maximum deviation from the average normalized peak value; *omomo* is a probe with asymmetric corrugations and alternation of oxide (*o*) and metal (*m*) bumps (adapted from [57]).

Variations in the peak value are above 90% and, more importantly, those in spot size exceed 280%. Such strong variations are due to the intrinsic directional nature of the asymmetries, with degradation in peak intensity and resolution as a consequence of misalignments from the preferential direction of the asymmetry.

7. Numerical investigation of probes with adirectional asymmetries under linearly polarized excitation

In order to reduce the sensitivity of the probe behaviour to the direction of the input polarization, we have introduced the concept of adirectional asymmetry. Two different implementations of this structural modification have been considered, both based on a spiral design: in fact, the spiral intrinsically fits the specification of lack of rotational and reflection symmetry and offers an interesting case study to investigate the feasibility of the concept of orientation-insensitive field localization [55,57]. First, the effects of a spiral corrugation on the outer metal surface of a fully metal-coated probe have been investigated. Then, another implementation based on azimuthal corrugations arranged in a spiral-like fashion will be discussed. A sketch of the two structures is shown in Figure 12.

The spiral corrugation is formed by joining a tapered helix-shaped 3D object (with circular cross-section of radius r) with two hemispherical terminations (having the same radius r). The spiral winding appears as a semicircular spiral corrugation and is placed between 150 nm and 750 nm along the z direction; the pitch along z is 300 nm. The spiral corrugation can take on the form of either a groove, i.e it is filled with air, or a bump, corresponding to metal filling; the effect of metal oxide filling has been analysed as well [57].

The azimuthal corrugations are formed by joining truncated toroids with hemispherical terminations as in paragraph 6. However, in order to create a spiral arrangement (and hence an adirectional asymmetry distributed over all spatial directions), the corrugations are shifted one with respect to the other [55].

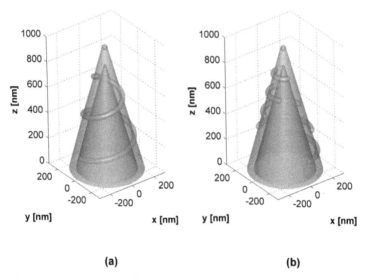

(a)   **(b)**

Figure 12. Sketch of the structures based on an adirectional asymmetry: (a) probe with spiral corrugation; (b) probe with spiral-arranged azimuthal corrugations.

7.1. Probe with spiral corrugation

The near field distributions obtained for a spiral metal corrugation with radius $r = 25$ nm under H and V polarized excitation are illustrated in Figure 13.

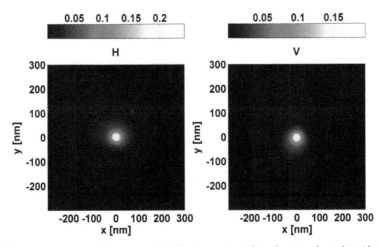

Figure 13. Normalized near-field intensity distributions at 10 nm from the apex of a probe with a spiral metal corrugation of radius r = 25 nm (adapted from [57]).

Strong field localization is observed for both the orthogonal input polarizations, with the creation of ultrasmall spots very similar in terms of both FWHM and peak value. As illustrated below, the intensity (that, in this case, is still 5 times lower than the one of the radial hot spot of the axisymmetric probe) can be adjusted by changing either the filling material or the radius of the corrugation. Similar near-field distributions have been observed upon variation in the chirality of the spiral winding (that is whether it wraps the tip in clockwise or counter-clockwise direction).

An analysis of the FWHM and the peak value (normalized to the peak achieved in the reference axisymmetric probe under radially polarized excitation) under different mutual orientations α of the direction of the linearly polarized excitation with respect to the asymmetry has been carried out to assess the properties of the adirectional asymmetry: according to Figure 14, α=0° corresponds to x linearly polarized excitation, with the two extremes of the spiral winding located along the x axis.

The most noticeable feature is that only minor fluctuations occur in the FWHM, which means that the spot size is almost insensitive to variations in the mutual orientation. We should remind that the FWHM is related to the eventual achievable resolution and, hence, its robustness with respect to variations in the orientation of the input polarization implies a substantial simplification in experimental applications. Although the peak value still depends on the mutual orientation, the maximum deviation from its average value calculated over all the mutual positions is below 20%, still tolerable if the average value were sufficiently high for detection. Note that, contrarily to the case of directional asymmetry where α=0° and α=90° represented the best and worst alignment, for the adirectional asymmetry different orientations are almost equivalent.

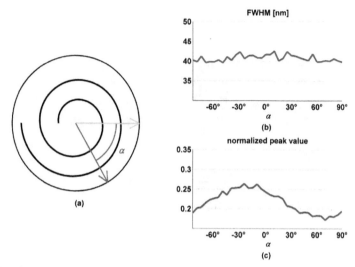

Figure 14. Behaviour of the probe with spiral corrugation for variable orientation of the input linearly polarized excitation with respect to the asymmetry: (a) schematic of the xy projection of the probe: the angle α is the one between the direction of the input linear polarization (magenta line) and the line along which the two extremes of the spiral are aligned (cyan line); (b) FWHM of the near-field distribution for the spiral metal corrugation under variable α; (c) peak value of the near-field distribution normalized to the one of the axisymmetric probe under radially polarized excitation for the spiral metal corrugation under variable α (adapted from [57]).

Similarly to the case of directional asymmetries, the FWHM and the peak value can be optimized by varying the characteristics of the spiral asymmetry. In particular, we changed the radius r of the spiral from 15 nm to 30 nm with a step of 5 nm. Moreover, we considered the case of replacement of the metal spiral bump with a spiral groove (air filling) or with an aluminium oxide spiral bump. In particular, for each structure based on a particular combination of filling material and radius r, the behaviour of the probe under all the possible mutual orientations of the linearly polarized excitation with respect to the spiral winding was examined, by calculating the average value of the peak ratio and the FWHM over all the mutual orientations and considering these values together with the maximum deviation from the corresponding average as figures of merit for comparison.

As for the directional asymmetries, overall better results (i.e. smaller spot size and higher peak values) have been obtained for stronger asymmetries due to larger r values. However, differently from the case of directional asymmetries, the data about the maximum deviation confirm that field localization is achieved irrespective of the mutual orientation between the linearly polarized mode and the asymmetry introduced in the structure. The deviation in FWHM dips as the radius increases, while the peak value shows an opposite trend with a change of the spiral radius: however, the maximum deviations are still acceptable, especially if combined with a rise in the average value occurring when the metal bump is substituted with an air groove or, even better, an aluminium oxide bump. The behaviour of the probe as a function of the filling material can be explained as a result of better coupling mechanisms between inner and outer SPPs, as also pointed out in paragraph 6.2.

7.2. Probe with corrugations arranged in a spiral-like fashion

The spiral corrugation can be replaced by a series of azimuthal corrugations shifted one with respect to the other to create a spiral arrangement, as visible in the 2D projection of Figure 15; the angle α between the direction of the input linearly polarized excitation and the angle bisector of the bottom corrugation is also defined. The parameters of this new structure can be varied to improve the field localization and enhancement for any orientation of the input linearly polarized excitation.

The effect of variations in the shift angle β (identical for any two consecutive corrugations) has been examined as well as the impact of variable radius and variable azimuthal extension γ of the single corrugations (which means the arc of circumference over which corrugations are spanned). According to the sketches in Figure 12 and Figure 15, the angle bisector of the bottom corrugation is along the x axis. The z spacing between consecutive corrugations, which can be either grooves or bumps, is 150 nm. We focused our attention on the structure based on alternation of grooves and bumps starting from the bottom corrugation, previously labelled as *amama*, because the studies reported in paragraph 6.2 for null shift angle β between consecutive corrugations showed that it resulted in the best performance.

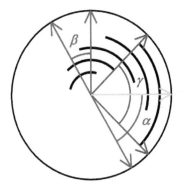

Figure 15. Schematic of the 2D projection of the probe with spiral-arranged corrugations and relevant parameters; the black arcs represent the corrugations; the angle α is the one between the direction of the input linear polarization (magenta line) and the angle bisector of the bottom corrugation (cyan line); the angle β is the shift angle measured between the angle bisectors of two consecutive corrugations, while the angle γ is the angular extension of each corrugation. The angle bisector of the bottom corrugation is along the x axis (adapted from [55]).

Figure 16 reports the square of the normalized near field distributions under H and V polarized excitation for a shift angle β of 45° (adirectional asymmetry), total azimuthal extension γ for each of the five corrugations equal to 160° and radius of 20 nm. The two distributions show close resemblance in both peak value and FWHM.

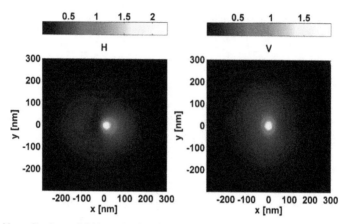

Figure 16. Normalized near-field intensity distributions at 10 nm from the apex of a probe with azimuthal corrugations arranged in a spiral-like fashion (β=45°; γ=160°; radius 20 nm) (adapted from [55]).

In order to further confirm that field localization is insensitive to the alignment of the input polarization, also for this asymmetry we studied the behaviour of the probe under variable orientation of the corrugations with respect to the input linear polarization, and, in particular, the FWHM and the peak value normalized to the one of the axisymmetric probe under radially polarized excitation.

Figure 17 shows deviations of the peak value from the average below 20% and those of the spot size less than 10% for the probe with adirectional asymmetry, which demonstrates the possibility to get superfocusing under arbitrarily oriented linearly polarized excitation. In particular, the ultimate achievable resolution is almost insensitive to the input polarization direction. Similar trends were observed by changing the chirality, i.e. the handedness, of the spiral-like arrangement.

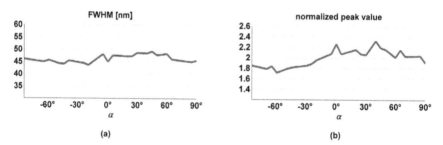

Figure 17. Behaviour of the probe with spiral-arranged corrugations for variable orientation α of the input linearly polarized excitation with respect to the asymmetry (β=45°; γ=160°; radius 20 nm): (a) FWHM of the near-field distribution for variable α; (b) peak value of the near-field distribution normalized to the one of the axisymmetric probe under radially polarized excitation for variable α (adapted from [55]).

The effect of some geometric parameters (more specifically the shift angle β, the angular extension γ and the radius of the corrugation) on the probe behaviour has been analysed as well. By increasing the shift angle from $0°$ to $45°$ (with a step of $7.5°$), a better confinement and an increase in the peak value has been observed, together with smaller variations in these parameters under variable orientation of the corrugations with respect to the input linearly polarized excitation. An increase in the angular extension and in the radius has also been shown to have an overall beneficial effect on probe behaviour.

8. Interaction of asymmetric probes with fluorescent molecules

Beyond the overall intensity distribution, also the analysis of the single vectorial components of the near-field distribution is of major interest and provides insight into the field localization mechanisms. A full vectorial picture of the electric field distribution close to the tip apex can be gained by scanning single fluorescent molecules with a SNOM probe. In the seminal work by Betzig and Chichester, the near-field distribution of aperture probes was mapped by fluorescent molecules, which act as point detectors for the components of the electric field aligned along their absorption dipole direction [75]. The interaction of single fluorescent molecules with aperture probes has been thoroughly investigated both experimentally and theoretically, not only as a tool for probe characterization, but also for its importance in biological as well as in material science [60,76-78].

For all these reasons, we have investigated the interaction of a single fluorescent molecule with an axisymmetric probe under radially polarized excitation and with an asymmetric probe under linearly polarized excitation in order to identify the similarities between the vectorial components of the near-field distributions close to the apex of the probes [79]. In most of the calculations, we considered the cut probe as an example of asymmetric probe due to the simplicity to fabricate this probe using FIB milling, as will be discussed in the next paragraph.

Figure 18 shows the quantities necessary to describe the dipole moment of the fluorescent molecules.

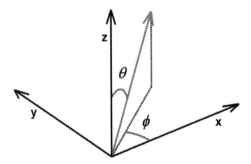

Figure 18. Representation of the orientation of the dipole (red vector) through the polar angle θ and the azimuthal angle ϕ.

The fluorescence intensity I emitted by a molecule is related to the square of the dot product of the local electric field E and the normalized absorption dipole moment p far from saturation [75]:

$$I \propto \left| E \cdot p \right|^2 \tag{4}$$

Therefore, the fluorescence intensity I depends on both the absolute value and the direction of the electric field, which shows the ability of a fluorescent molecule to provide a vectorial picture of the local electric field. The normalized absorption dipole moment describing the 3D dipole orientation is given by:

$$p = \begin{pmatrix} \sin\theta\cos\phi \\ \sin\theta\sin\phi \\ \cos\theta \end{pmatrix} \tag{5}$$

where θ and ϕ represent the polar and azimuthal angle, respectively (Figure 18). In our case the local field $E=(E_x,E_y,E_z)$ coincides with the electric field distribution on the xy plane at a distance z of 10 nm above the probe apex.

The interaction of an axisymmetric probe under radially polarized excitation with a fluorescent molecule with polar and azimuthal orientation varying from 0° to 90° (with a step of 30°) was considered. The resulting fluorescence intensity distributions calculated according to equation (4) on a 400 nm by 400 nm square area centered on the probe apex are shown in Figure 19, with an indication of the peak intensity for a molecule with a specific dipole orientation normalized to the peak of the total local electric field intensity $\left|E\right|^2$ (number reported in white). As visible, the radially polarized excitation gives rise to a strong longitudinal local field at the tip apex (aligned along the probe axis z and corresponding to $\theta=0°$), made up of a single hot spot, and weaker orthogonal transverse components (corresponding to $\theta=90°$ and $\phi=0°$ and $\theta=90°$ and $\phi=90°$, respectively) of almost equal magnitude appearing as two-lobed patterns.

A variation in the azimuthal angle of the fluorophore corresponds to almost identical intensity distributions, simply rotated according to ϕ, because of the axial symmetry of both the probe and the input excitation. On the contrary, a decrease in the peak intensity and a gradual transition from a single spot to a two-lobed distribution occurs upon a change of the polar angle θ from 0° to 90°, because of the overlap of the dipole absorption moment dominantly with the weaker transverse components of the local electric field rather than with the stronger longitudinal component.

The most remarkable feature is the dominance of a strong longitudinal field component, which can be beneficial for the imaging of molecules with a mainly longitudinal dipole moment. Moreover, molecules with identical polar angle and different azimuthal orientation give rise to similar fluorescent patterns apart from the rotation by ϕ, which allows the determination of the azimuthal component of the fluorophore dipole moment. On the contrary, using aperture probes, molecules with variable azimuthal angle result in strongly different patterns due to the significant differences in orthogonal transverse components close to the aperture, with variations in peak intensities by even two orders of magnitude [77].

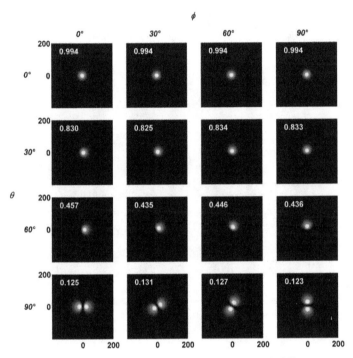

Figure 19. Simulated fluorescence intensity maps for single molecules with different orientations (as specified by the polar angle θ and the azimuthal angle ϕ) excited by an axisymmetric fully metal-coated probe under radially polarized excitation and located at 10 nm from the tip apex. All the maps are reported on the same colour scale, with each plot normalized to its peak intensity value (adapted from [79]).

A similar analysis has been run for a cut probe. Figure 20 shows the fluorescence intensity maps for a structure with a cut angle of 30° and a cut height of 766 nm under x linearly polarized excitation (i.e. oriented along the direction of the asymmetry; $\alpha=0°$). Also in this case, the dominance of the longitudinal component is clearly visible, with double-lobed orthogonal transverse components slightly different in magnitude. The close resemblance with the fluorescence distributions obtained for an axisymmetric probe under radially polarized excitation is significant, which makes the asymmetric structure eligible to replace axisymmetric probes in single fluorescent molecule studies, because the advantages pointed out for the axisymmetric probe can be obtained under an easier linearly polarized excitation.

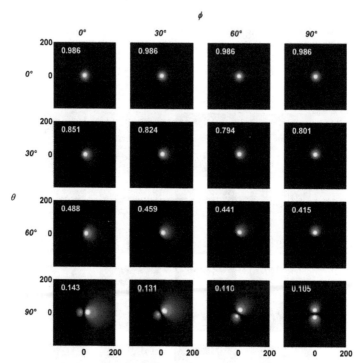

Figure 20. Simulated fluorescence intensity maps for single molecules with different orientations (as specified by the polar angle θ and the azimuthal angle ϕ) excited by a cut probe under linearly polarized excitation along the direction of the asymmetry (x) and located at 10 nm from the tip apex. All the maps are reported on the same colour scale, with each plot normalized to its peak intensity value (adapted from [79]).

The study of the interaction of different asymmetric probe structures with fluorescent molecules has revealed that all the asymmetric structures exhibit a highly confined strong longitudinal electric field component and weaker transverse components, like the axisymmetric probe under radially polarized excitation.

Moreover, the analysis of the intensity maps obtained for a fluorescent molecule with longitudinal dipole moment under variable α using either the previously considered cut probe or a probe with two-turn spiral corrugation in form of an aluminium bump with 25 nm radius (as in paragraph 7.1) has confirmed once again how adirectional asymmetries are more robust to variations in the direction of the input polarization compared to the directional ones. If the asymmetry lies along one specific spatial direction, alignment of the input linearly polarized excitation with respect to the asymmetry is still necessary. On the contrary, light funnelling to a highly confined mainly longitudinally polarized distribution independent of the orientation of the input linear polarization can be achieved by extending the asymmetry over all the spatial directions.

9. Hints on probe fabrication

In the introduction, we have briefly summarized the fundamental steps to create an aperture fiber probe. Our modified asymmetric probes can still benefit from this well developed technology. The overall fabrication process consists in the creation of the taper and subsequent metallization. Next, the desired modification is introduced by FIB milling. In our case, for first tests on the fabrication of our novel asymmetric probes, fiber-based SNOM tips have been bought and properly modified using a dual beam FIB/SEM (scanning electron microscope) system (FEI Strata DB 235). The system includes an ion beam and an electron beam column tilted one with respect to the other by 52°, allowing simultaneous nanostructuring and imaging of the probe.

FIB milling has been applied in the past to the fabrication of apertures in aperture SNOM probes, either to improve the quality of apertures created by some other method or to create an aperture in a fully metal-coated probe. Two different approaches have been used for this purpose, i.e. head-on drilling and slicing [80,81]. In the first approach the Ga⁺ beam is scanned from the top across the tip surface using a previously generated pattern (the ion beam is aligned along the tip axis); although in this way a better control over shape and size of the aperture can be achieved by choosing an appropriate milling pattern, deep drilling into the tip structure can occur and the ion beam needs to be guided to the center of the tip apex [80-82]. In the second approach, the metal-coated fiber probe is sliced off transversally, i.e. the ion beam is incident at 90° with respect to the probe axis; the size of the aperture can be controlled only indirectly by estimating the correct location for slicing and the size of the milled area; small beam sizes are used to remove thin slices of material corresponding to low ion currents [80,81,83-85].

The slicing approach, properly adapted, is suitable for the fabrication of our probes based on an oblique cut. However, differently from the case of conventional aperture probes, the angle of incidence of the ion beam with respect to the probe axis is no longer 90°, but needs to be chosen according to the desired cut angle. The definition of cut angle corresponds to the one given in paragraph 6: hence, a simple aperture probe would correspond to a cut angle of 0°.

The fiber probes used to fabricate and test the probe based on an oblique cut, bought from Lovalite, have been produced by heating and pulling and subsequent deposition of aluminium. For the fabrication, a step-by-step procedure was adopted in which a milling step consisting in scanning the ion beam over a small rectangular area to remove as little material as possible was followed by an imaging step in order to check the results of the previous nanostructuring step. To get a first knowledge of the aperture size and metal thickness, first simple aperture probes with diameters between about 70 nm and 200 nm were realized. Then cut probes were produced. Figure 21 shows an example of an aperture probe and of a probe with an oblique cut, with a cut angle of 50°.

FIB milling could also be applied for the creation of corrugations, even a spiral one: in fact, spiral lenses have been fabricated on planar structures [86,87]. In the wake of these encouraging results, we are confident that a spiral conical corrugation could be produced as well.

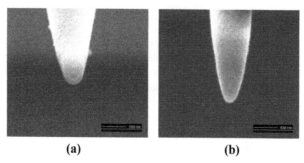

(a)   (b)

Figure 21. SEM image of: (a) aperture probe; (b) cut probe.

10. Conclusions

An extensive analysis of nanofocusing in asymmetric SNOM probe structures has been presented. The introduction of an asymmetry in an originally axisymmetric fully metal-coated probe structure has been shown to be effective in the achievement of field localization under linearly polarized excitation. In fact, in an apertureless axisymmetric structure, field localization, essential for high-resolution applications, can be attained only under radially polarized excitation. In this case, the surface plasmons excited by the radially polarized waveguide mode interfere constructively at the tip apex leading to the creation of an ultrasmall hot spot. On the contrary, those excited by a linearly polarized mode interfere destructively producing a broad, weak and asymmetric near-field distribution. However, the promising radially polarized excitation requires complicated injection procedures, extremely sensitive to misalignments.

Breaking the symmetry of the probe can help to avoid the destructive interference occurring under linearly polarized excitation allowing field localization at the tip apex under a more straightforward input excitation. Depending on the characteristics of the asymmetry, we have distinguished two different categories of asymmetry, which we called directional and adirectional asymmetries. Both the asymmetries break the axial symmetry of the original probe. However, the directional asymmetry consists in an asymmetry along one specific spatial direction; more specifically the structural modification exhibits a plane of symmetry. The adirectional asymmetry is devoid of reflection symmetry. In the first case, field localization is expected under linearly polarized excitation along the preferential spatial direction; in the second case, the field localization effect is almost insensitive to the direction of the input linearly polarized excitation. Hence, although both the categories of asymmetry obviate the need for a radially polarized excitation, adirectional asymmetries waive any requirement on the input polarization, with a further simplification of experimental procedures.

Probe design and modelling has been carried out adopting a numerical approach based on the finite element method. Two different implementations of directional asymmetries (a probe based on an oblique cut and one on asymmetric corrugations on the external metal coating) have been considered as well as two different configurations of the pioneering adirectional probe design (a probe with a spiral corrugation and one with azimuthal corrugations arranged in a spiral-like fashion). A 3D analysis of mode propagation has confirmed that properly tailored asymmetries can bring about field localization effects under linearly polarized excitation comparable to those achieved in an axisymmetric probe under radially polarized excitation. Moreover, the adirectional design guarantees more robustness against variations in the input polarization direction. Structural parameters can be optimized to get better near-field distributions in terms of FWHM and peak value. In general, probes with more extended asymmetries have shown to result in better overall performance.

The fabrication of the modified structures can benefit from well-established technologies for probe fabrication. In particular, the asymmetry itself can be introduced using FIB nanostructuring, as shown, for example, for the probe with an oblique cut. In this case, the slicing approach, already used for the creation of aperture probes, has been properly adapted for the creation of the novel probe.

In conclusion, these new promising probe concepts could result in significant headways towards high-resolution SNOM applications.

Author details

Valeria Lotito*
Electronics/Metrology Laboratory, Empa,
Swiss Federal Laboratories for Materials Science and Technology, Dübendorf, Switzerland
Department of Information Technology and Electrical Engineering, ETH Zurich, Zurich, Switzerland

Christian Hafner
Department of Information Technology and Electrical Engineering, ETH Zurich, Zurich, Switzerland

Urs Sennhauser
Electronics/Metrology Laboratory, Empa,
Swiss Federal Laboratories for Materials Science and Technology, Dübendorf, Switzerland

Gian-Luca Bona
Direction, Empa,
Swiss Federal Laboratories for Materials Science and Technology, Dübendorf, Switzerland
Department of Information Technology and Electrical Engineering, ETH Zurich, Zurich, Switzerland

Acknowledgement

The authors gratefully acknowledge the support of the Swiss National Science Foundation (project number 200021-115895) and the contribution of Dr. Konstantins Jefimovs (Empa) for probe nanostructuring.

---

*Corresponding Author

11. References

[1] Novotny L, Hecht B. Principles of Nano-Optics. New York: Cambridge University Press; 2006.

[2] Hecht B, Sick B, Wild U, Deckert V, Zenobi R, Martin O, Pohl D. Scanning near-field optical microscopy with aperture probes: fundamentals and applications. Journal of Chemical Physics 2000;112 7761-7774.

[3] Lewis A, Isaacson M, Harootunian A, Muray A. Development of a 500 Å spatial resolution light microscope. Ultramicroscopy 1984;13 227-231.

[4] Pohl DW, Denk W, Lanz M. Optical stethoscopy: Image recording with resolution $\lambda/20$. Applied Physics Letters 1984;44 651-653.

[5] Ambrosio A, Fenwick O, Cacialli F, Micheletto R, Kawakami Y, Gucciardi PG, Kang DJ, Allegrini M. Shape dependent thermal effects in apertured fiber probes for scanning near-field optical microscopy. Journal of Applied Physics 2006;99 084303.

[6] La Rosa AH, Yakobson BI, Hallen HD. Origins and effects of thermal processes on near-field optical probes. Applied Physics Letters 1995;67(18) 2597-2599.

[7] Gucciardi PG, Colocci M, Labardi M, Allegrini M. Thermal-expansion effects in near-field optical microscopy fiber probes induced by laser light absorption. Applied Physics Letters 1999;75(21) 3408-3410.

[8] Novotny L, Hafner C. Light propagation in a cylindrical waveguide with a complex metallic dielectric function. Physical Review E 1994;50 4094-4106.

[9] Antosiewicz TJ, Marciniak M, Szoplik T. On SNOM resolution improvement. In: Sibilia C, Benson TM, Marciniak M, Szoplik T (eds.) Photonic crystals: physics and technology. Milan: Springer; 2008 p. 217–238.

[10] Saiki T, Matsuda K. Near-field optical fiber probe optimized for illumination-collection hybrid mode operation. Applied Physics Letters 1999;74(19) 2773-2775.

[11] Yatsui T, Kourogi M, Ohtsu M. Increasing throughput of a near-field optical fiber probe over 1000 times by the use of a triple-tapered structure. Applied Physics Letters 1998;73(15) 2090-2092.

[12] Mononobe S, Saiki T, Suzuki T, Koshihara S, Ohtsu M. Fabrication of a triple-tapered probe for near-field optical microscopy in UV region based on selective chemical etching of a multistep index fiber. Optics Communications 1998;146 45-48.

[13] Nakamura H, Sato T, Kambe H, Sawada K, Saiki T. Design and optimization of tapered structure of near-field fibre probe based on finite-difference time-domain simulation. Journal of Microscopy 2001;202(1) 50-52.

[14] Yatsui T, Kourogi M, Ohtsu M. Highly efficient excitation of optical near-field on an apertured fiber probe with an asymmetric structure. Applied Physics Letters 1997;71(13) 1756-1758.

[15] Garcia-Parajo M, Tate T, Chen Y. Gold-coated parabolic tapers for scanning near-field optical microscopy: fabrication and optimisation. Ultramicroscopy 1995;61 155-163.

[16] Arslanov NM. The optimal form of the scanning near-field optical microscopy probe with subwavelength aperture. Journal of Optics A: Pure and Applied Optics 2006;8(3) 338-344.

[17] Bakunov MI, Bodrov SB, Hangyo M. Intermode conversion in a near-field optical fiber probe. Journal of Applied Physics 2004;96(4) 1775-1780.

[18] Antosiewicz TJ, Szoplik T. Corrugated metal-coated tapered tip for scanning near-field optical microscope. Optics Express 2007;15(17) 10920–10928.

[19] Wang Y, Wang Y-Y, Zhang X. Plasmonic nanograting tip design for high power throughput near-field scanning aperture probe. Optics Express 2010;18(13) 14004-14011.

[20] Matteo JA, Fromm DP, Yuen Y, Schuck PJ, Moerner WE, Hesselink L. Spectral analysis of strongly enhanced visible light transmission through single C-shaped nanoapertures. Applied Physics Letters 2004;85(4) 648-650.

[21] Jin EX, Xu X. Finite-difference time-domain studies on optical transmission through planar nano-apertures in a metal film. Japanese Journal of Applied Physics 2004;43(1) 407-417.

[22] Danzebrink HU, Dziomba T, Sulzbach T, Ohlsson O, Lehrer C, Frey L. Nano-slit probes for near-field optical microscopy fabricated by focused ion beams. Journal of Microscopy 1999;194(2/3) 335-339.

[23] Butter JYP, Hecht B. Aperture scanning near-field optical microscopy and spectroscopy of single terrylene molecules at 1.8 K. Nanotechnology 2006;17 1547-1550.

[24] Tanaka K, Tanaka M. Optimized computer-aided design of I-shaped subwavelength aperture for high intensity and small spot size. Optics Communications 2004;233 231-244.

[25] Minh PN, Ono T, Tanaka S, Esashi M. Spatial distribution and polarization dependence of the optical near-field in a silicon microfabricated probe. Journal of Microscopy 2001;202(1) 28-33.

[26] Bortchagovsky E, Colas des Francs G, Naber A, Fischer UC. On the optimum form of an aperture for a confinement of the optically excited electric near field. Journal of Microscopy 2008;229(2) 223-227.

[27] Li Z-B, Zhou W-Y, Kong X-T, Tian J-G. Polarization dependence and independence of near-field enhancement through a subwavelength circle hole. Optics Express 2010;18(6) 5854-5860.

[28] Li Z-B, Zhou W-Y, Kong X-T, Tian J-G. Near-field enhancement through a single subwavelength aperture with gaps inside. Plasmonics 2011;6(1) 149-154.

[29] Bouhelier A, Novotny L., Near field optical excitation and detection of surface plasmons. In Brongersma ML, Kik PG (eds.) Surface Plasmon Nanophotonics, Dordrecht The Netherlands: Springer; 2007 p. 139-153.

[30] Hartschuh A. Tip-enhanced near-field optical microscopy. Angewandte Chemie International Edition 2008;47 8178-8191.

[31] Stockman MI. Nanofocusing of optical energy in tapered plasmonic waveguides. Physical Review Letters 2004;93(13) 137404.

[32] Babadjanyan AJ, Margaryan NL, Nerkararyan KV. Superfocusing of surface polaritons in the conical structure. Journal of Applied Physics 2000;87(8) 3785-3788.

[33] Kurihara K, Otomo A, Syouji A, Takahara J, Suzuki K, Yokoyama S. Superfocusing modes of surface plasmon polaritons in conical geometry based on the quasi-separation of variables approach. Journal of Physics A – Mathematical and theoretical. 2007;40(41) 12479–12503.

[34] Vogel MW, Gramotnev DK. Optimization of plasmon nano-focusing in tapered metal rods. Journal of Nanophotonics 2008;2 1-17.

[35] Ruppin R. Effect of non-locality on nanofocusing of surface plasmon field intensity in a conical tip. Physics Letters A 2005;340(1-4) 299-302.

[36] Baida FI, Belkhir A. Superfocusing and light confinement by surface plasmon excitation through radially polarized beam. Plasmonics 2009;4(1) 51-59.

[37] Maier SA, Andrews SR, Martìn-Moreno L, Garcìa-Vidal FJ. Terahertz surface plasmon-polariton propagation and focusing on periodically corrugated metal wires. Physical Review Letters 2006;97 176805.

[38] Lee JS, Han S, Shirdel J, Koo S, Sadiq D, Lienau C, Park N. Superfocusing of electric or magnetic fields using conical metal tips: effect of mode symmetry on the plasmon excitation method. Optics Express 2011;19(13) 12342-12347.

[39] Ropers C, Neacsu CC, Elsaesser T, Albrecht M, Raschke MB, Lienau C. Grating-coupling of surface plasmons onto metallic tips: a nanoconfined light source. Nano Letters 2007;7(9) 2784-2788.

[40] Ding W, Andrews SR, Maier SA. Internal excitation and superfocusing of surface plasmon polaritons on a silver-coated optical fiber tip. Physical Review A 2007;75 063822.

[41] Janunts NA, Baghdasaryan KS, Nerkararyan KV, Hecht B. Excitation and superfocusing of surface plasmon polaritons on a silver-coated optical fiber tip. Optical Communications 2005;253(1-3) 118–124.

[42] Nerkararyan K, Abrahamyan T, Janunts E, Khachatryan R, Harutyunyan S. Excitation and propagation of surface plasmon polaritons on the gold covered conical tip. Physics Letters A 2006;350(1-2) 147-149.

[43] Babayan AE, Nerkararyan KV. The strong localization of surface plasmon polariton on a metal-coated tip of optical fiber. Ultramicroscopy 2007;107(12) 1136-1140.

[44] Nerkararyan KV, Hakhoumian AA, Babayan AE. Terahertz surface plasmon-polariton superfocusing in coaxial cone semiconductor structures. Plasmonics 2008;3 27-31.

[45] Abrahamyan T, Nerkararyan K. Surface plasmon resonance on vicinity of gold-coated fiber tip. Physics Letters A 2007;364(6) 494–496.

[46] Bouhelier A, Renger J, Beversluis MR, Novotny L. Plasmon-coupled tip-enhanced near-field optical microscopy. Journal of Microscopy 2003;210(3) 220-224.

[47] Chen W, Zhan Q. Numerical study of an apertureless near field scanning optical microscope probe under radial polarization illumination. Optics Express 2007;15(7) 4106-4111.

[48] Liu L, He S. Design of metal-cladded near-field fiber probes with a dispersive body-of-revolution finite-difference time-domain method. Applied Optics 2005;44(17) 3429-3437.

[49] Vaccaro L, Aeschimann L, Staufer U, Herzig HP, Dändliker R. Propagation of the electromagnetic field in fully coated near-field optical probes. Applied Physics Letters 2003;83(3) 584-586.

[50] Tortora P, Descrovi E, Aeschimann L, Vaccaro L, Herzig HP, Dändliker R. Selective coupling of HE11 and TM01 modes into microfabricated fully metal-coated quartz probes. Ultramicroscopy 2007;107(2-3) 158-165.

[51] Lotito V, Sennhauser U, Hafner C. Effects of asymmetric surface corrugations on fully metal-coated scanning near field optical microscopy tips Optics Express 2010;18(8) 8722-8734.

[52] Nakagawa W, Vaccaro L, Herzig HP. Analysis of mode coupling due to spherical defects in ideal fully metal-coated scanning near-field optical microscopy probes. Journal of the Optical Society of America A 2006;23(5) 1096-1105.

[53] Frey HG, Keilmann F, Kriele A, Guckenberger R. Enhancing the resolution of scanning near-field optical microscopy by a metal tip grown on an aperture probe. Applied Physics Letters 2002;81(26) 5030-5032.

[54] Taminiau TH, Segerink FB, Moerland RJ, Kuipers L(K), van Hulst NF. Near field driving of a optical monopole antenna. Journal of Optics A: Pure and Applied Optics 2007;9 S315-S321.

[55] Quong MC, Elezzabi AY. Offset-apertured near-field scanning optical microscope probes. Optics Express 2007;15(16) 10163-10174.

[56] Lotito V, Sennhauser U, Hafner C, Bona G-L. Fully metal-coated scanning near-field optical microscopy probes with spiral corrugations for superfocusing under arbitrarily oriented linearly polarized excitation. Plasmonics 2011;6 327-336.

[57] Lotito V, Sennhauser U, Hafner C, Bona G-L. A novel nanostructured scanning near field optical microscopy probe based on an adirectional asymmetry. Journal of Computational and Theoretical Nanoscience 2012;9 486-494.

[58] Roberts A. Electromagnetic theory of diffraction by a circular aperture in a thick, perfectly conducting screen. Journal of the Optical Society of America A 1987;4 1970-1983.

[59] Roberts A. Near zone fields behind circular apertures in thick, perfectly conducting screens. Journal of Applied Physics 1989;65 2896-2899.

[60] Dunn RC. Near-field scanning optical microscopy. Chemical Reviews 1999;99 2891-2927.

[61] Issa NA, Guckenberger R. Optical nanofocusing on tapered metallic waveguides. Plasmonics 2007;2 31-37.

[62] Ohtsu M. Progress in Nano-electro-optics III Industrial applications and dynamics of the nano-optical system. Berlin Heidelberg: Springer; 2005.

[63] Girard C, Dereux A. Near field optics theories Report on Progress in Physics 1996;59 657-699.

[64] Girard C. Near fields in nanostructures. Report on Progress in Physics 2005;68 1883-1933.

[65] Moar PN, Love JD, Laudoceur F, Cahill LW. Waveguide analysis of heat-drawn and chemically etched probe tips for scanning near-field optical microscopy. Applied Optics 2006;45(25) 6442-6456.

[66] Novotny L, Pohl DW, Hecht B. Scanning near field optical probe with ultrasmall size. Optics Letters 1995;20 970-972.

[67] Wang XQ, Wu S-F, Jian G-S, Pan S. The advantages of a pyramidal probe tip entirely coated with a thin metal film for SNOM. Physics Letters A 2003;319(5-6) 514-517.

[68] Frey HG, Bolwien C, Brandenburg A, Ros R, Anselmetti D. Optimized apertureless optical near-field probes with 15 nm optical resolution. Nanotechnology 2006;17(13) 3105-3110.

[69] Nakagawa W, Vaccaro L, Herzig HP, Hafner C. Polarization mode coupling due to metal-layer modifications in apertureless near-field scanning optical microscopy probes. Journal of Computational and Theoretical Nanoscience 2007;4 692–703.

[70] Bondeson A, Rylander T, Ingelström P. Computational Electromagnetics. New York: Springer; 2005.

[71] Rao SS. The finite element method in engineering. Burlington MA USA: Butterworth-Heinemann; 1999.

[72] Lotito V, Sennhauser U, Hafner C. Finite element analysis of asymmetric scanning near field optical microscopy probes. Journal of Computational and Theoretical Nanoscience 2010;7 1596–1609.

[73] Janunts NA, Nerkararyan KV. Modulation of light radiation during input into a waveguide by resonance excitation of surface plasmons. Applied Physics Letters 2001;79(3) 299–301.

[74] Lotito V, Sennhauser U, Hafner C. Numerical analysis of novel asymmetric SNOM tips. PIERS Online 2011;7(4) 394-400.

[75] Betzig E, Chichester RJ. Single molecules observed by near-field scanning optical microscopy Science 1993;262(5138) 1422-1425.

[76] Hollars W, Dunn RC. Probing single molecule orientations in model lipid membranes with near-field scanning optical microscopy. Journal of Chemical Physics 2000;112(18) 7822-7830.

[77] Veerman JA, M. Garcia-Parajo F, Kuipers L, van Hulst NF. Single molecule mapping of the optical field distribution of probes for near field microscopy. Journal of Microscopy 1999;194(2-3), 477-482.

[78] Moerland RJ, van Hulst NF, Gersen H, Kuipers L. Probing the negative permittivity perfect lens using near-field optics and single molecule detection. Optics Express 200513(5) 1604-1614.

[79] Lotito V, Sennhauser U, Hafner C, Bona G-L. Interaction of an asymmetric scanning near field optical microscopy probe with fluorescent molecules. Progress In Electromagnetics Research 2011;121 281-299.

[80] Heinzelmann H, Freyland JM, Eckert R, Huser T, Schürmann G, Noell W, Staufer U, de Rooij NF. Towards better scanning near-field optical microscopy probes – progress and new developments. Journal of Microscopy 1999;194(2-3) 365-368.

[81] Muranishi M, Sato K, Osaka S, Kikukawa A, Shintani T, Ito K. Control of aperture size of optical probes for scanning near-field optical microscopy using focused ion beam technology. Japanese Journal of Applied Physics 1997;36(7B), L942-L944

[82] Lacoste T, Huser T, Prioll R, Heinzelmann H. Contrast enhancement using polarization-modulation scanning near-field optical microscopy (PM-SNOM). Ultramicroscopy 1998;71 333–340.

[83] Veerman JA, Otter AM, Kuipers L, van Hulst NF. High definition aperture probes for near-field optical microscopy fabricated by focused ion beam milling. Applied Physics Letters 1998;72 115–3117.

[84] Pilevar S, Edinger K, Atia A, Smolyaninov I, Davis C. Focused ion-beam fabrication of fiber probes with well-defined apertures for use in near-field scanning optical microscopy. Applied Physics Letters 1998;72 3133–3135.

[85] Zhang Y, Dhawan A, Vo-Dinh T. Design and fabrication of fiber-optic nanoprobes for optical sensing. Nanoscale Research Letters 2011;6(18) 6 pp.

[86] Yang S, Chen W, Nelson RL, Zhan Q. Miniature circular polarization analyzer with spiral plasmonic lens. Opics Letters 2009;34(20) 3047-3049.

[87] Chen W, Abeysinghe DC, Nelson RL, Zhan Q. Experimental confirmation of miniature spiral plasmonic lens as a circular polarization analyzer. Nano Letters 2010;10 2075–2079.

Chapter 7

Propagating Surface Plasmons and Dispersion Relations for Nanoscale Multilayer Metallic-Dielectric Films

Henrique T. M. C. M. Baltar, Krystyna Drozdowicz-Tomsia and Ewa M. Goldys

Additional information is available at the end of the chapter

1. Introduction

The study of propagating surface plasmons (PSPs) is an important aspect of the understanding of the interaction between light and metallic surfaces. One of the key concepts related to PSPs, is the dispersion relation. This relation is the basis for understanding of coupling of light to PSPs, by using special approaches to match the wavevector. Moreover, it can be used to predict the matching of localised surface plasmons (LSPs) and PSPs to achieve highly enhanced electromagnetic field and/or tailored transmission.

This chapter is concerned with a detailed analysis of propagating surface plasmons (PSPs), and the calculations of the dispersion relations in nanoscale multilayer metallic-dielectric films, starting from the fundamental Maxwell's equations. Furthermore, we discuss the symmetric IMI (insulator-metal-insulator) and MIM (metal-insulator-metal) geometries, as well as their asymmetric variants. We will also describe the PSP in the IIMI (insulator-insulator-metal-insulator) geometry.

There is a vast literature on the subject, however, numerous authors assume the materials as lossless by using the Drude's model without damping. We initially model the dielectric function of the metals in our multilayer structures by using the lossless Drude's model. Then, we model the metals by applying complex values of permittivities acquired experimentally. The Drude's model without damping, with its simplicity, provides a basic understanding, as all the calculated quantities are purely real or imaginary numbers. Nevertheless, real structures behave differently from the predictions of this simple model. The dispersion relation changes markedly, when complex permittivities are used, and the changes are more pronounced around the surface plasmon frequency. This leads to wavevector limitations and to the existence of a region of anomalous dispersion, called quasi-bound mode [4]. Therefore, it is necessary to extend the analysis to include complex permittivity, which has been tabulated for the most common materials.

There are some previous studies on the behaviour of the IMI and MIM structures using complex permittivity, but the discussions are restrained to symmetric structures and/or do not show the entire dispersion curve [4, 5, 32]. In many cases, such symmetries are not applicable to the real experiments. For instance, assays with analytes deposited over a metallic layer on a substrate are, in this sense, an asymmetric problem.

Other experimentally important structure is the IIMI. Such structure is used for surface-enhanced fluorescence (SEF). A dielectric separation layer between metal and fluorophore can be introduced to maximise the fluorescence by reducing the possibility of quenching of fluorophores in direct contact with metal surfaces [6]. To the best of our knowledge, there are no published studies of the dispersion relation for such type of asymmetric structures. These new calculations are presented in this work.

2. Permittivity of metals

The Drude's model, taking into account only free electrons, is applicable to metals and leads to the following expression for the relative permittivity [16, 23]:

$$\varepsilon_r = 1 - \frac{\omega_p^2}{\omega^2 + i\Gamma\omega} \tag{1}$$

Here, ω_p is called the plasma frequency, and the damping constant Γ is the electron scattering rate (the inverse of the collision time for the conduction electrons). When a metal is assumed lossless (very long time between collisions, $\Gamma \to 0$), the permittivity becomes real:

$$\text{Drude's permittivity without damping:} \quad \varepsilon_r = 1 - \frac{\omega_p^2}{\omega^2} \tag{2}$$

In this study, we also model the metal permittivity by the tabulated values of silver from Lynch and Hunter [20]. In the case of asymmetric MIM geometry, we will also use the tabulated permittivity of gold from Johnson and Christy [13]. In Figure 1, we plotted the real and imaginary parts of the permittivities of silver and gold. For the Drude's model without damping, we used the plasma frequency of 13.1 Prad (2.08 PHz, $\lambda_p = 144$ nm) for silver and 13.4 Prad (2.13 PHz, $\lambda_p = 141$ nm) for gold.

3. Theoretical framework

3.1. Source-free Maxwell's equations in time-harmonic regime

The Maxwell's equations hold for any arbitrary time-dependence of the electric field. For materials in a linear regime we can consider these equations for each frequency component separately [1]. The time-dependent electric or magnetic fields can be written as the real part of a complex field, denoted by F:

$$\vec{F}(\vec{r}, t) = \vec{F}_0(\vec{r}) \cos\left(\vec{k} \cdot \vec{r} - \omega t + phase\right) = Re\left[\vec{F}(\vec{r}) e^{-i\omega t}\right] \tag{3}$$

---

[1] The non-linear regime is when the material properties such as permittivity and permeability depend on frequency and intensity of the field, and will not be considered here.

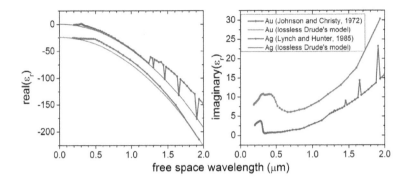

Figure 1. Permittivities of metals. Silver is modelled by the Drude's model without damping ($\omega_p = 13.1$ Prad, $\lambda_p = 144$ nm) and the tabulated values of Lynch and Hunter (1985) [20]. Gold is modelled by the Drude's model without damping ($\omega_p = 13.4$ Prad, $\lambda_p = 141$ nm) and the tabulated values of Johnson and Christy (1972) [13]. For easiness of visualisation, the real part of gold permittivity was shifted -25 units, and the imaginary part, $+5$ units.

Thus a field can be described by a vector phasor $\vec{F}(\vec{r})$, which contains information on the magnitude, direction and phase [3].

In plasmonics, it is common to deal with structures without any external current sources and/or charges. In this situation, we apply the source-free macroscopic Maxwell's equations, in which the displacement charges and currents are fully incorporated in the permittivities and permeabilities of the materials.

In the time-harmonic regime, the source-free macroscopic Maxwell's equations can be written as [3]

$$\begin{aligned}
\text{Gauss' law of electricity:} \quad & \vec{\nabla} \cdot \vec{D}(\vec{r}) = 0 \\
\text{Gauss' law of magnetism:} \quad & \vec{\nabla} \cdot \vec{B}(\vec{r}) = 0 \\
\text{Faraday's law of induction:} \quad & \vec{\nabla} \times \vec{E}(\vec{r}) = i\omega\vec{B}(\vec{r}) \\
\text{Ampère's circuital law:} \quad & \vec{\nabla} \times \vec{H}(\vec{r}) = -i\omega\vec{D}(\vec{r})
\end{aligned} \tag{4}$$

3.2. Boundary conditions

The source-free Maxwell's equations in their integral form applied to the interface between two different media (1 and 2) lead to four boundary conditions [10, 29]:

$$\begin{aligned}
\hat{n} \cdot \left(\vec{D}_2 - \vec{D}_1\right) &= 0 \\
\hat{n} \cdot \left(\vec{B}_2 - \vec{B}_1\right) &= 0 \\
\hat{n} \times \left(\vec{E}_2 - \vec{E}_1\right) &= 0 \\
\hat{n} \times \left(\vec{H}_2 - \vec{H}_1\right) &= 0
\end{aligned} \tag{5}$$

These boundary conditions describe the continuity of the components of the fields perpendicular and parallel to the interface. Moreover, the momentum parallel to the interface is also continuous [12]. In such case, one can write the boundary conditions as

$$\vec{D}_{2\perp} = \vec{D}_{1\perp} \tag{6a}$$
$$\vec{B}_{2\perp} = \vec{B}_{1\perp} \tag{6b}$$
$$\vec{E}_{2\parallel} = \vec{E}_{1\parallel} \tag{6c}$$
$$\vec{H}_{2\parallel} = \vec{H}_{1\parallel} \tag{6d}$$
$$\vec{k}_{2\parallel} = \vec{k}_{1\parallel} \tag{6e}$$

3.3. Electromagnetic waves

The Maxwell's equations represent a coupled system of differential equations. With some algebraic manipulation, it is possible to transform these four equations (4) into two uncoupled ones, representing the homogeneous vector wave equations for the electric and magnetic fields. In the time-harmonic regime, these wave equations become the homogeneous vector Helmholtz' equations [3]:

$$\nabla^2 \vec{E} + k^2 \vec{E} = 0$$
$$\nabla^2 \vec{H} + k^2 \vec{H} = 0 \tag{7}$$

In which

$$k = \frac{\omega}{v} = nk_0 = \sqrt{\mu \varepsilon} k_0 \tag{8}$$

is the wavenumber in a medium with index of refraction $n = \sqrt{\mu \varepsilon}$, and $k_0 = \omega/c$ is the free-space wavenumber.

3.4. Light polarisation

Due to the linearity of the homogeneous Helmholtz' equations, we can divide the problem of a wave incident to a planar surface into two parts: for an s- and for a p-polarised wave[2]. These waves are depicted in Figure 2, in which we define the plane of incidence as the x-z plane.

In these conditions, for a plane wave propagating in the x-z plane, the magnitudes of the phasor fields are dependent on the coordinates x and z, but constant along the y direction. In order to find the surface plasmons — which are waves bound to the interfaces —, we can write the magnitudes of the electric and magnetic phasor fields as

$$\vec{E}(\vec{r}) = \vec{E}(x,z) = \vec{E}(z) e^{i\beta x}$$
$$\vec{H}(\vec{r}) = \vec{H}(x,z) = \vec{H}(z) e^{i\beta x} \tag{9}$$

Where β is the propagation constant, in the x direction. As the parallel wavevector to the interface must be constant (Eq. 6e), we have

$$\beta = k_{x,1} = k_{x,2} \tag{10}$$

---

[2] The terms s- and p-polarised are the initials of the German words: senkrecht (perpendicular) and parallel. In this nomenclature, s or p means that the electric field is perpendicular (s) or parallel (p) to the plane of incidence [23]. The s-polarisation is also called TE (transverse electric); and the p-polarisation, TM (transverse magnetic) [3, 29].

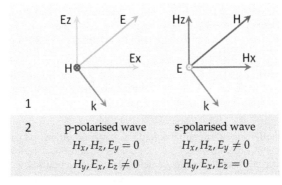

Figure 2. Interface between two media (1 and 2), showing the polarisation vectors of a p-polarised (left) and an s-polarised waves (right). The plane of incidence coincides with the page. A p-polarised wave (TM) is characterised by the magnetic field perpendicular to the plane of incidence; while an s-polarised (TE) wave is characterised by the electric field perpendicular to the plane of incidence.

In this case, the differential operators with respect to the coordinates x and y can be written as

$$\frac{\partial}{\partial x} = i\beta$$
$$\frac{\partial}{\partial y} = 0 \tag{11}$$

This makes it possible to express the divergence and Laplacian of any of the fields as

$$\vec{\nabla} \times \vec{F}(x,z) = \begin{vmatrix} e_x & e_y & e_z \\ i\beta & 0 & \frac{\partial}{\partial z} \\ F_x & F_y & F_z \end{vmatrix} = \left(-\frac{\partial F_y}{\partial z}\right)\hat{e}_x + \left(\frac{\partial F_x}{\partial z} - i\beta F_z\right)\hat{e}_y + (i\beta F_y)\hat{e}_z \tag{12a}$$

$$\nabla^2 \vec{F}(x,z) = \frac{\partial^2 \vec{F}}{\partial z^2}(x,z) - \beta^2 \vec{F}(x,z) \tag{12b}$$

3.4.1. Revisiting Helmholtz' equations and boundary conditions

The Laplacian in Equation 12b and the fields in Equation 9 can be used to simplify the Helmholtz' equations (7) to one dimension [21]:

$$\frac{\partial^2 \vec{E}}{\partial z^2}(z) - \gamma_z^2 \vec{E}(z) = 0 \tag{13a}$$

$$\frac{\partial^2 \vec{H}}{\partial z^2}(z) - \gamma_z^2 \vec{H}(z) = 0 \tag{13b}$$

With

$$\gamma_j^2 = -k_{z,j}^2 = \beta^2 - \varepsilon_j k_0^2 \tag{14}$$

$k_{z,j}$ is the wavevector along the z axis in the medium j ($= 1, 2$). We define the propagation constant γ as

$$\gamma = ik_z \tag{15}$$

This definition is commonly done in transmission-line theory [3], and it is also used in most of the literature about PSPs. We note that, when dealing with a theoretical situation of lossless materials, working with γ implies in dealing with real numbers instead of imaginary.

It is sufficient to solve one of the Helmholtz' equation for the polarisation whose field (electric or magnetic) is perpendicular to the plane of incidence (Figure 2). In other words, for the s-polarised wave, whose electric field has only the y-component, we solve Equation 13a; while for the p-polarised wave, whose magnetic field has only the y-component, we solve Equation 13b.

The Helmholtz' equations are differential equations of second order, and each one needs two boundary conditions in order to be solved. Equations 6c and 6d, describing the continuity of the parallel electric and parallel magnetic fields, provide one boundary condition for each of the Helmholtz' equations. Below, we provide one additional boundary condition for each field (electric and magnetic).

3.4.2. s-polarisation

Expanding the divergence (Equation 12a) in the Faraday's law (Equation 4) for an s-polarised wave (Figure 2), we get

$$\left(-\frac{\partial E_y}{\partial z}\right)\hat{e}_x + \left(i\beta E_y\right)\hat{e}_z = \left(i\omega B_x\right)\hat{e}_x + \left(i\omega B_z\right)\hat{e}_z \tag{16}$$

- **z component**: The continuity of the perpendicular $\vec{B}$ field (Equation 6b), B_z, implies the continuity of βE_y. As β is continuous (Equation 10), this condition is equivalent to the previous condition of continuity of E_y (Equation 6c).

- **x component**: For homogeneous isotropic materials, $\vec{B} = \mu\vec{H}$. Therefore,

$$i\omega H_x = -\frac{1}{\mu}\frac{\partial E_y}{\partial z} \tag{17}$$

This equation, together with the continuity of the parallel magnetic field (Equation 6d), leads to the continuity of $\frac{1}{\mu}\frac{\partial E_y}{\partial z}$.

In summary, the conditions for the s-polarised wave are:

$$\text{Continuity conditions for s-polarised wave:} \quad \begin{aligned} E_{y,1} &= E_{y,2} \\ \frac{1}{\mu_1}\frac{\partial E_{y,1}}{\partial z} &= \frac{1}{\mu_2}\frac{\partial E_{y,2}}{\partial z} \end{aligned} \tag{18}$$

3.4.3. p-polarisation

For the p-polarised wave, we use the same approach as for the s-polarised one. Expanding the divergence (Equation 12a), in the Ampère's law (Equation 4), we obtain

$$\left(-\frac{\partial H_y}{\partial z}\right)\hat{e}_x + \left(i\beta H_y\right)\hat{e}_z = -\left(i\omega D_x\right)\hat{e}_x - \left(i\omega D_z\right)\hat{e}_z \tag{19}$$

- **z component**: The continuity of the perpendicular $\vec{D}$ field (Equation 6a), D_z, implies the continuity of βH_y. As β is continuous (Equation 10), this condition is equivalent to the previous condition of continuity of H_y (Equation 6d).

- **x component**: For homogeneous isotropic materials, $\vec{D} = \varepsilon \vec{E}$. Therefore,

$$i\omega E_x = \frac{1}{\varepsilon}\frac{\partial H_y}{\partial z}. \tag{20}$$

This equation together with the continuity of the parallel electric field (Equation 6c) leads to the continuity of $\frac{1}{\varepsilon}\frac{\partial H_y}{\partial z}$ [2].

In summary, the conditions for the p-polarised wave are:

$$\text{Continuity conditions for p-polarised wave:} \quad \begin{array}{c} H_{y,1} = H_{y,2} \\ \dfrac{1}{\varepsilon_1}\dfrac{\partial H_{y,1}}{\partial z} = \dfrac{1}{\varepsilon_2}\dfrac{\partial H_{y,2}}{\partial z} \end{array} \tag{21}$$

4. PSP at a planar interface

Now we discuss the conditions satisfied by a wave propagating at a planar interface between two half-spaces. So as to solve the problem of an s-polarised wave propagating at an interface, we will use Equation 13a, and the two boundary conditions stated in Equations 18. For the p-polarised wave, we will use Equation 13b and the boundary conditions in Equations 21.

4.1. s-polarised wave

In order to solve Equation 13a, we postulate a solution of a certain form, and verify that it indeed satisfies our equation and boundary conditions. We are looking for a wave bound to the interface, that vanishes away from it. With this in mind, and inspecting Equation 13a, we postulate our solution to be

$$\vec{E}(z) = \begin{cases} E_1 e^{-\gamma_1 z}\hat{e}_y, & z > 0 \\ E_2 e^{\gamma_2 z}\hat{e}_y, & z < 0 \end{cases} \tag{22}$$

with the real parts of γ_j positive in order to have a field decaying away from the interface. Applying the boundary conditions to this ansatz, we find

$$\begin{array}{c} E_1 = E_2 \\ \dfrac{\gamma_1}{\mu_1}E_1 = -\dfrac{\gamma_2}{\mu_2}E_2 \end{array} \tag{23}$$

These conditions are equivalent to [27]

$$\frac{\gamma_1}{\mu_1} = -\frac{\gamma_2}{\mu_2} \tag{24}$$

For the materials found in nature, the permeability at optical frequencies is close to unity[3] [27]. This leads to $\gamma_1 = -\gamma_2$, what is impossible as the real components of both γ_1 and γ_2 are

---

[3] The relative permeability of unity in the optical regime for most of the materials implies that the index of refraction is $n = \sqrt{\varepsilon\mu} = \sqrt{\varepsilon}$.

positive. However, magnetic surface plasmons can be achieved by the use of metamaterials [24], for which the magnetic permeability can be engineered to be negative at frequencies higher than in natural materials. For example, ref. [19] shows magnetic plasmon propagation at infra-red along a chain of split-ring resonators.

4.2. p-polarised wave

Similarly to the case of an s-polarised wave, we propose the following solution for the Equation 13b:

$$\vec{H}(z) = \begin{cases} H_1 e^{-\gamma_1 z}\hat{e}_y, & z > 0 \\ H_2 e^{\gamma_2 z}\hat{e}_y, & z < 0 \end{cases} \tag{25}$$

Applying the boundary conditions:

$$\begin{aligned} H_1 &= H_2 \,(= H_0) \\ \frac{\gamma_1}{\varepsilon_1}H_1 &= -\frac{\gamma_2}{\varepsilon_2}H_2 \end{aligned} \tag{26}$$

These conditions lead to [7, 21, 23, 27, 28, 34]

$$\frac{\gamma_1}{\varepsilon_1} = -\frac{\gamma_2}{\varepsilon_2} \tag{27}$$

This condition combined with Equation 14 and the index of refraction[3] $n = \sqrt{\varepsilon}$ leads to the dispersion relation[4] [4, 7, 21, 23, 34]:

$$\text{Dispersion relation for half-spaces:} \quad \beta^2 = k_0^2 \frac{\varepsilon_1 \varepsilon_2}{\varepsilon_1 + \varepsilon_2} \tag{28}$$

Substituting this β^2 in Equation 14, one obtains the normal component of the wavevector [4, 23]:

$$k_{z,j}^2 = -\gamma_j^2 = k_0^2 \frac{\varepsilon_j^2}{\varepsilon_1 + \varepsilon_2} \tag{29}$$

4.2.1. Conditions for PSPs

In this subsection, we limit our discussion on the conditions for the existence of PSPs to lossless materials. Subsequently, we will discuss the effect of complex permittivities on β and γ. In lossless materials, all permittivities as well as the propagation constants (Equations 28 and 29) are real. Without loss of generality, one can define the surface wave propagating at the direction $+x$. Hence, β is positive (Equation 28). 1) This requires that $\varepsilon_1\varepsilon_2$ and $\varepsilon_1 + \varepsilon_2$ are both positive or both negative. Furthermore, as we are looking for the modes bound to the interface, we want the field propagating at the boundary, but not along the z axis. Therefore, γ must be positive real (Equation 25) (or $k_{z,1}$ and $k_{z,2}$ must be positive imaginary, Equation 15), leading to evanescent, exponentially decaying fields away from the interface. 2) This requirement is achievable only if $\varepsilon_1 + \varepsilon_2$ is negative (Equation 29). As a result, the conditions for the existence of PSPs are [23]

---

[4] The dispersion relation is an equation that relates the wavevector along the direction of propagation and the frequency.

$$\text{Conditions for PSPs:} \quad \begin{matrix} \varepsilon_1\varepsilon_2 < 0 \\ \varepsilon_1 + \varepsilon_2 < 0 \end{matrix} \tag{30}$$

The first condition states that one of the permittivities must be positive, and the other one, negative. The second condition, indicates that the absolute value of the negative permittivity must be higher than the positive one [23].

As already seen in Figure 1 and Equation 2, lossless Drude's metals have negative permittivities for $\omega < \omega_p$ (or $\lambda > \lambda_p$). The absolute values of these permittivities increase with the wavelength, exceeding the values found in dielectrics. Semiconductors [26] and graphene [9, 25] have also been used for their plasmonic properties in the terahertz range. Although in graphene there is no strict surface plasmon, due to its two-dimensional geometry, which practically prevents transverse oscillations of electrons [11]. As metals are the most common materials used for PSPs, a planar interface that supports bound waves will henceforth be called insulator-metal (IM) geometry.

Other important consideration to take into account, when calculating the wavevectors, is the position of β in relation to the light line. If β is at right of the light line ($\beta^2 > \varepsilon_j k_0^2$) (Equation 14) in the medium j, γ_j is real ($k_{z,j}$ is imaginary). If β is at left ($\beta^2 < \varepsilon_j k_0^2$), γ is imaginary ($k_{z,j}$ is real), leading to the propagation of the wave away from the interface [2].

4.2.2. Complex β and γ in lossy materials

When the permittivity of any medium in the structure is complex, the normal and perpendicular wavevectors components are complex. The magnetic field (Equations 9 and 25) can be re-written in the following form:

$$\vec{H}\left(\vec{r}\right) = \vec{H}\left(z\right)e^{i\beta x} = \begin{cases} H_0 e^{i\beta x}e^{-\gamma_1 z}\hat{e}_y, & z > 0 \\ H_0 e^{i\beta x}e^{\gamma_2 z}\hat{e}_y, & z < 0 \end{cases} \tag{31}$$

At the interface ($z = 0$), this converges to

$$\vec{H}\left(x,y,0\right) = H_0 e^{i\beta x}\hat{e}_y = H_0 e^{iRe(\beta)x}e^{-Im(\beta)x}\hat{e}_y \tag{32}$$

Looking at this equation, we remind that the meaning of the (positive) real part of β is that the wave propagates on the interface towards $+x$. If the imaginary part of β is negative, the field increases exponentially (Equation 32), which is un-physical in our situation. Consequently, the imaginary part of β must be positive, leading to an exponential decay with x. Lossless materials ($Im\left(\beta = 0\right)$) imply an endless propagation of the PSP along the interface. One can define the propagation length as the distance the wave travels along the interface until its energy decays to e^{-1} (≈ 0.368) of its original value. As the energy is proportional to the square of the field ($energy \propto |H|^2 \propto e^{-2Im(\beta)x}$), the propagation length is [7, 21, 34]

$$propagation\ length \equiv L_{sp} = \frac{1}{2Im\left(\beta\right)} \tag{33}$$

Now we analyse the component of the wavevector that is perpendicular to the interface, γ_j. For a fixed x, Equation 31 can be re-written as

$$\vec{H}\left(constant,y,z\right) = H_0 e^{i\beta x} \times \begin{cases} e^{-Re(\gamma_1)z}e^{-iIm(\gamma_1)z}\hat{e}_y, & z > 0 \\ e^{Re(\gamma_2)z}e^{iIm(\gamma_2)z}\hat{e}_y, & z < 0 \end{cases} \tag{34}$$

For bound modes, we remind that $Re(\gamma_j)$ must be positive (or $Im(k_{z,j})$ must be negative). Equation 14 tells γ has two solutions in anti-phase in the complex plane. One must choose, therefore, the solutions in the first and fourth quadrants.

4.2.3. Dispersion curve

Applying the lossless Drude's model (Equation 2) to the dispersion relation for the IM geometry (Equation 28), one obtains the result presented in Figure 3A. In this figure, we plotted the solution for two different dielectrics: free-space and a material with $n = 1.5$ throughout the entire spectral range. The dispersion is divided into two branches with a region between them where β is imaginary. In this region, called plasmon bandgap [4], the wave is evanescent and it reflects back to the dielectric. For the Drude's model without damping (Equation 2), the index of refraction of a metal is real when the angular frequency is larger than the plasma frequency ($\omega > \omega_p$). Therefore, the material is transparent and the wave propagates freely [12, 28]. This radiation mode is called Brewster mode [28]. Under the plasmon bandgap, there is a region limited by what is called the surface plasmon frequency, the frequency in which β diverges. It occurs when the permittivity of the metal (ε_m) cancels that one of the dielectric (ε_d) (Equation 28) [4, 21, 34]:

$$\varepsilon_m = -\varepsilon_d \Rightarrow \omega_{sp} = \frac{\omega_p}{\sqrt{1 + \varepsilon_d}} \tag{35}$$

The higher the index of refraction of the dielectric is, the lower the surface plasmon frequency is. From this frequency down, light propagates in a bound mode. For a very high wavevector[5], the electric field has an electrostatic character, as the wavelength along the interface ($= 2\pi/\beta$) tends to zero. Likewise, the group velocity ($d\omega/dk$) [21] and the decay lengths into the metal and dielectric ($= 1/Re(\gamma_j)$) tend to zero. This high confinement of light at the interface indicates a true surface mode, called Fano mode [28]. For very low frequencies, the dispersion converges to the light line, in a regime called Sommerfeld-Zenneck waves [21, 28].

Using the complex experimental values of permittivity, the dispersion relation (Equation 28) is shown is Figure 3B. As the metal is no longer modelled as lossless, β is complex, leading to a finite propagation length (Equation 33), drawn in Figure 3C. Here, the wavenumber does not diverge as previously. It bends backwards filling the region previously called plasmon bandgap and connects to the Brewster mode. This region of anomalous dispersion (negative phase velocities $= d\omega/dk$) is called quasi-bound mode [4].

In the infra-red, the imaginary part of the permittivity of silver is very small compared to the real part. As a consequence, the propagation length is around 10–1000 µm. In the region of the quasi-bound mode, the imaginary component reaches values comparable to the real component (up to 0.83 of the real part for n=1, up to 1.9 for n=1.5), leading to a decrease in the propagation length down to 10–100 nm.

5. Physical interpretation

An electromagnetic wave inside a material induces polarisation in the medium. The coupled excitation resulting from the interaction between an incident electromagnetic wave and the

---

[5] In surface plasmons, the wavevector is higher than it would be in the dielectric, leading to a wavelength shorter than the one in the dielectric. This has applications in super-resolution light microscopy [15, 33].

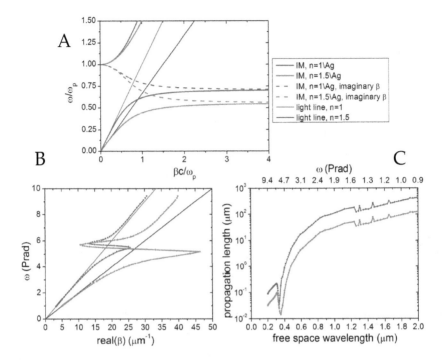

Figure 3. Dispersion curve and propagation length (Equation 28) for PSPs on an interface between half-spaces of a metal and insulators with refractive indices of 1 or 1.5. A) Drude's metal without damping. The imaginary β in the bandgap is also shown. B) Dispersion curve and C) propagation length for tabulated values of the metal (silver) permittivity.

material polarisation is called polariton [22, 31]. The way the coupling occurs is responsible for permittivity, permeability and light velocity being different from those ones in vacuum [22]. When the polariton is restricted to a small region around an interface, it is called surface polariton [31]. There are diverse types of polaritons according to the induced excitation, such as phonon-polariton [8], exciton-polariton [30], plasmon-polariton etc. The latter is the interest in the context of PSPs.

Plasmon is technically a quantum of charge density oscillation in a plasma [18]. The plasmons can be divided into two main types: bulk plasmons — arising from fluctuations of the free charge density inside the material which propagate as a longitudinally-polarised charge-density wave [18] — and surface plasmons (SPs) — which propagate along the interfaces of materials under specific circumstances. The SPs can be subdivided in localised surface plasmons (LSPs) and propagating surface plasmons (PSPs). In the former, the charge oscillations are confined to a nanostructure; while in the latter, the charge oscillation propagates relatively large distances, from tens of nanometers to hundreds of micrometers (Figure 3C).

We showed in the previous section that electromagnetic waves can propagate along the interfaces between dielectrics and metals. In response to an incidence electric field, the electrons of the metal induce an opposing field determined by the dielectric function of the metal. The fact that in metals the charge density is confined to a small region close to the surface, leads to high enhancement of the electric field, which can be as high as about 100 times[6] [27].

6. Exciting PSPs

As demonstrated previously, the wavenumber of a PSP is higher than the wavenumber in the dielectric (Figure 3A and B). Hence, light incident at an IM interface typically cannot excite PSPs. In order to do so, it is necessary to artificially increase the wavenumber of the light, so that the wavevectors in both materials can be matched. Some of the techniques to excite PSPs include the Otto and Kretschmann configurations, and the application of a grating or other periodic structure.

The Otto and Kretschmann configurations [7, 23, 27, 34] relies on the proximity of a dielectric with a higher index of refraction than the dielectric of the interface where the PSP is being excited. As the wavenumber is directly proportional to the index of refraction (Equation 8), one can use the evanescent waves that arise from a total internal reflection on a high-refractive index material to excite PSPs. In the Otto configuration, the reflectance comes from the side of the interface dielectric; while in the Kretschmann one, the evanescent waves come across a thin metal layer to excite the interface on the other side.

Diffraction gratings provide light with an additional momentum on the grating axis due to the spatial periodicity, allowing the coupling of light to PSPs [7, 27]. For a one-dimensional lattice, the additional momentum is given by

$$\Delta k = m \frac{2\pi}{a} \tag{36}$$

Where a is the period of the grating, and m is an integer.

7. PSP at a thin layer

The actual geometry of a thin layer (medium 2) between two media (1 and 3) is presented in Figure 4A. As previously discussed for an IM geometry, only p-polarised wave can excite PSP in naturally-occurring materials. This statement is also valid for other geometries. In a p-polarised wave, there is an electric field perpendicular to the surface (Figure 2). As D_z is continuous (Equation 6a), E_z changes, resulting in the creation of charge density at the interface. In an s-polarised wave, the electric field has no component parallel to the surface, therefore it is continuous and it does not produce charge density at the surface [14]. For modes bound to the interface in p-polarisation, we postulate the magnetic field — stronger at the interface and exponentially decaying away from it — as

$$H_y(z) = \begin{cases} H_1 e^{-\gamma_1 z}, & z > \frac{a}{2} \\ H_2^- e^{-\gamma_2 z} + H_2^+ e^{\gamma_2 z}, & \frac{a}{2} > z > -\frac{a}{2} \\ H_3 e^{\gamma_3 z}, & -\frac{a}{2} > z \end{cases} \tag{37}$$

Applying the continuity conditions (Equations 21) to the ansatz given in Equation 37, we

---

[6] This enhancement of the electric field may be applied for surface-enhanced spectroscopies, such as surface-enhanced Raman spectroscopy (SERS), surface-enhanced fluorescence (SEF) [17] and surface-enhanced infra-red absorption (SEIRA) [1].

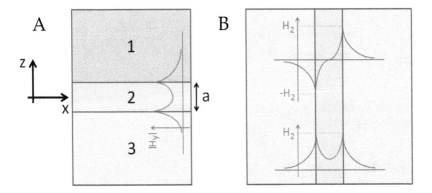

Figure 4. Planar thin layer geometry. A) Geometry of a planar thin layer surrounded by two half-spaces. It also shows the absolute value of the magnetic field expected for modes bound to the interfaces. B) The field magnitudes for the symmetric and anti-symmetric PSPs in a symmetric thin layer.

obtain four equations, two for each boundary.

- Boundary 1–2 ($z = a/2$)

$$H_1 e^{-\gamma_1 a/2} = H_2^- e^{-\gamma_2 a/2} + H_2^+ e^{\gamma_2 a/2} \tag{38}$$

$$\frac{\gamma_1}{\varepsilon_1} H_1 e^{-\gamma_1 a/2} = \frac{\gamma_2}{\varepsilon_2} H_2^- e^{-\gamma_2 a/2} - \frac{\gamma_2}{\varepsilon_2} H_2^+ e^{\gamma_2 a/2} \tag{39}$$

- Boundary 2–3 ($z = -a/2$)

$$H_3 e^{-\gamma_3 a/2} = H_2^- e^{\gamma_2 a/2} + H_2^+ e^{-\gamma_2 a/2} \tag{40}$$

$$\frac{\gamma_3}{\varepsilon_3} H_3 e^{-\gamma_3 a/2} = -\frac{\gamma_2}{\varepsilon_2} H_2^- e^{\gamma_2 a/2} + \frac{\gamma_2}{\varepsilon_2} H_2^+ e^{-\gamma_2 a/2} \tag{41}$$

To simplify the notation, we define

$$R_i = \frac{\gamma_i}{\varepsilon_i} \tag{42}$$

Substituting $H_1 e^{-\gamma_1 a/2}$ in Equation 39 by 38, $H_3 e^{-\gamma_3 a/2}$ in 41 by 40, and multiplying by $e^{\gamma_2 a/2}$, we obtain

$$R_1 \left(H_2^- + H_2^+ e^{\gamma_2 a} \right) = R_2 \left(H_2^- - H_2^+ e^{\gamma_2 a} \right) \tag{43}$$

$$R_3 \left(H_2^+ + H_2^- e^{\gamma_2 a} \right) = R_2 \left(H_2^+ - H_2^- e^{\gamma_2 a} \right) \tag{44}$$

The original system of four variables is now reduced to two: H_2^+ and H_2^-. Re-arranging the terms, we can obtain the following relationship in a matrix form:

$$\begin{pmatrix} e^{\gamma_2 a} \left(R_1 + R_2 \right) & R_1 - R_2 \\ R_3 - R_2 & e^{\gamma_2 a} \left(R_2 + R_3 \right) \end{pmatrix} \begin{pmatrix} H_2^+ \\ H_2^- \end{pmatrix} = \begin{pmatrix} 0 \\ 0 \end{pmatrix} \tag{45}$$

This is a homogeneous system of linear equations. Its non-zero solutions occur when the determinant of the matrix of coefficients is zero, resulting in [21]

Dispersion relation for thin layer: $e^{2\gamma_2 a} = \dfrac{R_1 - R_2}{R_1 + R_2}\dfrac{R_3 - R_2}{R_3 + R_2}$

$\qquad\qquad$ (46)

γ_2 and R_i are given by Equations 14 and 42, respectively.

7.1. Symmetric structures

Suppose that media 1 and 3 are composed of the same material. In this case, the dispersion relation (Equation 46) becomes

$$e^{2\gamma_2 a} = \left(\frac{R_1 - R_2}{R_1 + R_2}\right)^2 \qquad (47)$$

This equation has two roots [28]:

$$e^{\gamma_2 a} = \pm\frac{R_1 - R_2}{R_1 + R_2} \qquad (48)$$

Which can be re-arranged to

$$\frac{R_1}{R_2} = -\frac{e^{\gamma_2 a} \pm 1}{e^{\gamma_2 a} \mp 1} = -\frac{e^{\gamma_2 a/2} \pm e^{-\gamma_2 a/2}}{e^{\gamma_2 a/2} \mp e^{-\gamma_2 a/2}} \qquad (49)$$

Or [21]

$$\frac{R_1}{R_2} = \begin{cases} -coth\left(\frac{\gamma_2 a}{2}\right) \\ -tanh\left(\frac{\gamma_2 a}{2}\right) \end{cases} \qquad (50)$$

Returning to Equation 43 and re-arranging, we obtain

$$\frac{R_1}{R_2} = \frac{H_2^- - H_2^+ e^{\gamma_2 a}}{H_2^- + H_2^+ e^{\gamma_2 a}} = -\frac{e^{\gamma_2 a/2} - e^{-\gamma_2 a/2}\frac{H_2^-}{H_2^+}}{e^{\gamma_2 a/2} + e^{-\gamma_2 a/2}\frac{H_2^-}{H_2^+}} \qquad (51)$$

Comparing Equations 50 and 51 leads to [4]

Dispersion relation for symmetric thin layer:
$$\begin{aligned}\frac{R1}{R2} &= -coth\left(\frac{\gamma_2 a}{2}\right) \Rightarrow H_2^+ = -H_2^- \\ \frac{R1}{R2} &= -tanh\left(\frac{\gamma_2 a}{2}\right) \Rightarrow H_2^+ = H_2^-\end{aligned} \qquad (52)$$

Therefore, in symmetric structures, the PSPs has two branches: a symmetric (or even) and an anti-symmetric (or odd). A sketch of the field magnitude is shown in Figure 4B.

There are two possibilities for a symmetric structure: IMI (insulator-metal-insulator) and MIM (metal-insulator-metal). We will show in sequence the dispersion curves for the symmetric IMI, symmetric MIM and their respective asymmetric variants.

7.2. Dispersion curves for a thin layer

Applying lossless Drude's metal for the dispersion relation for symmetric IMI (Equations 52), we obtain the curves depicted in Figure 5A and B. In 5A, the metal is immersed in free space, while in 5B, it is immersed in a dielectric with $n = 1.5$. The dispersion curves of the light lines in the dielectrics and of the respective IM geometries are also plotted for comparison. In the thin layer case, the dispersion relationship splits in two modes. The even mode appears below the dispersion of the IM, while the odd mode appears above it. At very low frequencies, both of the modes converge to the light line, as does the IM dispersion. For frequencies close to ω_{sp}, the wavevectors of both modes tend to infinity. The bulk mode appears only for the even mode, very similar to the one of the IM dispersion, but with higher wavevectors. The thinner is the layer, the farther the modes are from the IM dispersion. For thicker layers, the two modes converge to the IM dispersion [32], as expected due to lower interactions between the two interfaces. We note that more than one solution is possible for the odd mode, as shown in the inset in Figure 5A. Indeed, for the 25 nm layer, there were up to three solutions for each frequency.

For the symmetric MIM structures, the dispersion relation for a lossless Drude's metal is plotted in Figure 5C. The shape changes markedly from the IM dispersion. Now, the odd mode appears below the dispersion of the IM. This is the opposite of the result for the IMI geometry. Furthermore, the solution for the bulk mode appears only from the equation (52) of the odd mode. The even mode partially coincides with the plasmon bandgap, producing a quasi-bound mode. Close to ω_{sp}, the wavevectors of both modes tend to infinity. For very low frequencies, the odd mode does not tend to the light line as in the IMI structure. In addition, as expected, the thinner the layer is, the farther its dispersion is from the IM dispersion. The inset shows that, for the 50 nm layer, the even mode has two solutions for a range of frequencies.

The dispersion curve (Equation 46) for a 25 nm asymmetric IMI, is presented in Figure 5D. The mode that would be equivalent to the even mode in the symmetric variant appears under the IM dispersion with the high refractive-index dielectric. For very low frequencies, this mode converges to the light line of the same dielectric. The mode equivalent to the odd mode in the symmetric variant appears above the IM dispersion with the low refractive-index dielectric. Close to the respective ω_{sp}, the wavevector of this mode tends to infinity. For lower frequencies, there is a cut off. This mode disappears when it reaches the light line of the high refractive-index dielectric (see arrow in Figure 5D).

For a 25 nm asymmetric MIM (Figure 5E), the mode under the IM dispersions tends to infinity in the lower ω_{sp} (silver); while the mode above the IM dispersions tend to infinity in the higher ω_{sp} (gold). Silver has a lower plasma frequency than gold. Therefore, the IM dispersion of silver and vacuum remains under the IM dispersion of gold and vacuum. Between the ω_{sp} of the two half-spaces, there is a gap. In addition, the bulk mode also presents a cut-off frequency above the plasma frequencies of the two metals, when it reaches the IM dispersion of silver and vacuum (see arrow in Figure 5E).

The dispersion curves of the symmetric variants of IMI and MIM modelling the metals by the experimental values of permittivity are presented in Figure 6. As metals are modelled as lossy, the propagation lengths are also plotted. For the symmetric IMI with 25 and 50 nm lossy metal layer (Figure 6A and B), the odd modes are present in the entire frequency range. However, the even mode wavevectors of the IMI enters the bandgap region and decreases down to zero and re-appear in the bulk region. In the region of real surface plasmons, its wavevector is greater than the one of the odd mode. The odd mode, with smaller wavevectors, has propagation lengths longer than the lengths of the IM (Figure 6B), which — in its turn — are typically longer than the propagation length of the even mode. The even mode may have a longer propagation length only in the bulk mode, region not important for PSPs.

For the symmetric MIM structure with complex permittivities (Figure 6C and D), the even mode does not present a solution. At ω_{sp}, the odd mode wavevector is greater than the one in the IM dispersion. Its propagation length is shorter than the one of the IM case.

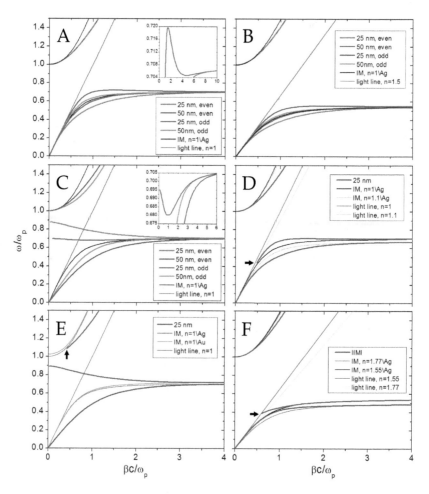

Figure 5. Dispersion relations for diverse geometries and lossless Drude's metals. A) Symmetric IMI: 25 and 50 nm metal layer immersed in free space. B) Similar to the previous geometry, but the metal is immersed in a dielectric with n=1.5. C) Symmetric MIM: 25 and 50 nm free-space layer immersed in metal. D) Asymmetric IMI: 25 nm metal layer surrounded by free space and other dielectric with n=1.1. E) Asymmetric MIM: 25 nm free space surrounded by silver and gold. F) IIMI: free space, 15 nm layer of a dielectric with n=1.77, 45 nm metal layer and a dielectric with n=1.55. The ω_p on the axes is the one of the silver. The insets in A and C show selected curves with different axes limits.

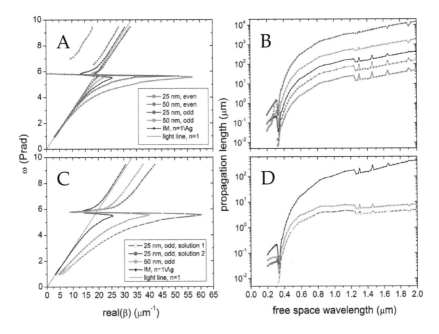

Figure 6. Dispersion curves and propagation lengths for symmetric geometries modelling the metals by
the tabulated permittivities. A, B) IMI: 25 and 50 nm silver layer immersed in free space. C, D) MIM: 25
and 50 nm free-space layer immersed in silver.

The dispersion curves and propagation lengths of the asymmetric variants of IMI and MIM
for tabulated permittivities are presented in Figure 7. For the IMI structure (Figure 7 A and
B), instead of two solutions, three solutions appear. One of them (solution 3) has very short
propagation length throughout the spectrum, and very low wavevector, except in the bulk
region. Therefore, this mode does not produce PSP. The mode with great wavevector (solution
1), as the one in the symmetric case, enters the region of quasi-bond mode and decreases down
to zero, with a bandgap, appearing again in the bulk region. Another mode (solution 2), with
wavevectors close to the light line ($n = 1.1$), presented regions of cut-off in low frequencies as
well as in the bulk region.

The MIM structure (Figure 7C and D) shows only one solution in the whole frequency
spectrum. It is interesting to note that this solution presents two regions of PSPs, and
consequently two quasi-bond mode regions. These regions corresponds to the ω_{sp} of the IM
dispersion with silver and the ω_{sp} of the IM dispersion with gold. The propagation lengths
are shorter than in any of the IM cases, with silver or with gold.

8. PSP at an IIMI structure

The geometry of an IIMI structure is depicted in Figure 8A. The structure consists of a thin
metallic layer (m) deposited on a dielectric substrate (4), and covered by a thin dielectric layer
(2) with free space on the top of the whole structure (1). The solution for the magnetic field of

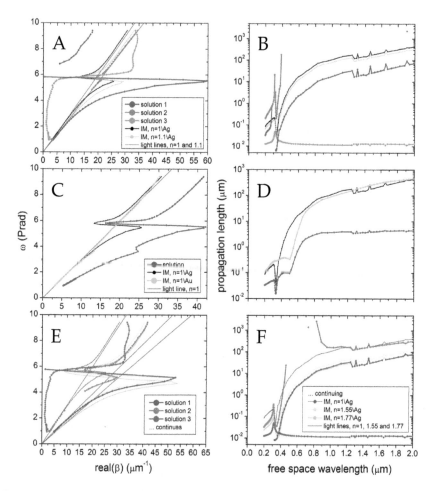

Figure 7. Dispersion relations and propagation lengths for asymmetric geometries modelling the metals by the tabulated permittivities. A, B) IMI: 25 nm silver layer surrounded by free space and other dielectric with n=1.1. C, D) MIM: 25 nm free-space layer surrounded by silver and gold. E, F) IIMI: free space, 15 nm layer of a dielectric with n=1.77, 45 nm silver layer and a dielectric with n=1.55.

the p-polarised wave decaying away from the interfaces is postulated as

$$
H_y = e^{i\beta x} \times \begin{cases} A_1 e^{-\gamma_1 z}, & a+b < z \\ A_2^- e^{-\gamma_2 z} + A_2^+ e^{\gamma_2 z}, & a < z < a+b \\ A_m^- e^{-\gamma_m z} + A_m^+ e^{\gamma_m z}, & 0 < z < a \\ A_4 e^{\gamma_4 z}, & z < 0 \end{cases} \tag{53}
$$

One should not expect to have a PSP at the 1–2 interface, as both materials are dielectrics. Nevertheless, it is necessary to take into account evanescent waves decaying away from this interface. Otherwise, the IIMI structure would simply behave like an IMI.

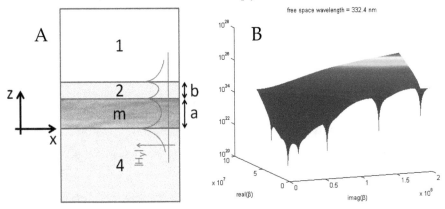

Figure 8. A) Geometry of the IIMI structure. The metal layer is located on the top of a dielectric half-space, and it is covered by a layer of other dielectric with free space above. A sketch of the absolute value of the magnetic field, as expected by modes bound to the interfaces, is also presented. B) Example of absolute values surface of the determinant of the matrix **C** (Equation 63) in the space of complex β for $\lambda = 332.4\,nm$. The minima which converge to zero represent solutions for the dispersion equation of the IIMI structure (Equation 64).

Applying the continuity conditions (Equation 21), we obtain six equations, two for each boundary.

- Boundary 1–2 ($z = a + b$)

$$A_1 e^{-\gamma_1(a+b)} = A_2^- e^{-\gamma_2(a+b)} + A_2^+ e^{\gamma_2(a+b)} \tag{54}$$

$$\frac{\gamma_1}{\varepsilon_1} A_1 e^{-\gamma_1(a+b)} = \frac{\gamma_2}{\varepsilon_2} A_2^- e^{-\gamma_2(a+b)} - \frac{\gamma_2}{\varepsilon_2} A_2^+ e^{\gamma_2(a+b)} \tag{55}$$

- Boundary 2–m ($z = a$)

$$A_2^- e^{-\gamma_2 a} + A_2^+ e^{\gamma_2 a} = A_m^- e^{-\gamma_m a} + A_m^+ e^{\gamma_m a} \tag{56}$$

$$\frac{\gamma_2}{\varepsilon_2} A_2^- e^{-\gamma_2 a} - \frac{\gamma_2}{\varepsilon_2} A_2^+ e^{\gamma_2 a} = \frac{\gamma_m}{\varepsilon_m} A_m^- e^{-\gamma_m a} - \frac{\gamma_m}{\varepsilon_m} A_m^+ e^{\gamma_m a} \tag{57}$$

- Boundary m–4 ($z = 0$)

$$A_m^- + A_m^+ = A_4 \tag{58}$$

$$-\frac{\gamma_m}{\varepsilon_m} A_m^- + \frac{\gamma_m}{\varepsilon_m} A_m^+ = \frac{\gamma_4}{\varepsilon_4} A_4 \tag{59}$$

Substituting $A_1 e^{-\gamma_1(a+b)}$ in Equation 55 by Equation 54 yields

$$\frac{\gamma_1}{\varepsilon_1} \left(A_2^- e^{-\gamma_2(a+b)} + A_2^+ e^{\gamma_2(a+b)} \right) = \frac{\gamma_2}{\varepsilon_2} \left(A_2^- e^{-\gamma_2(a+b)} - A_2^+ e^{\gamma_2(a+b)} \right) \tag{60}$$

Substituting A_4 in 59 by 58,

$$\frac{\gamma_m}{\varepsilon_m}\left(-A_m^- + A_m^+\right) = \frac{\gamma_4}{\varepsilon_4}\left(A_m^- + A_m^+\right) \tag{61}$$

With Equations 60 and 61, we reduced the system to 4 variables: A_2^-, A_2^+, A_m^- and A_m^+. These two equations in addition to Equations 56 and 57 can be written in a matrix form:

$$\mathbf{C}\begin{pmatrix} A_2^- \\ A_2^+ \\ A_2 \\ A_m^- \\ A_m^+ \end{pmatrix} = \begin{pmatrix} 0 \\ 0 \\ 0 \\ 0 \\ 0 \end{pmatrix} \tag{62}$$

In which $\mathbf{C}$ is a 4×4 matrix of coefficients:

$$\mathbf{C} = \begin{pmatrix} R_1 e^{-\gamma_2(a+b)} - R_2 e^{-\gamma_2(a+b)} & R_1 e^{\gamma_2(a+b)} + R_2 e^{\gamma_2(a+b)} & 0 & 0 \\ 0 & 0 & R_4 + R_m & R_4 - R_m \\ e^{-\gamma_2 a} & e^{\gamma_2 a} & -e^{-\gamma_m a} & -e^{\gamma_m a} \\ R_2 e^{-\gamma_2 a} & -R_2 e^{\gamma_2 a} & -R_m e^{-\gamma_m a} & R_m e^{\gamma_m a} \end{pmatrix} \tag{63}$$

γ_i and R_i are given by Equations 14 and 42, respectively. For a homogeneous system of linear equations, the non-zero solutions occur when the determinant of the matrix of coefficients is zero:

$$\text{Dispersion relation for IIMI:} \quad |\mathbf{C}| = 0 \tag{64}$$

8.1. Dispersion curves for an IIMI structure

The dispersion curve of this last structure for a Drude's model without damping is presented in Figure 5F. For comparison, the light lines of the dielectrics adjacent to the metal are drawn, as well as the half-spaces of the same dielectrics with the metal. This structure behaves close to an asymmetric IMI. It presents one mode, under the lower half-spaces dispersion. The wavevector of this mode tends to infinity at ω_{sp} of the lower half-spaces dispersion. The mode above the two half-spaces dispersion tends to infinity at ω_{sp} of the other half-spaces dispersion. For lower frequencies, the same mode reaches the low refractive-index light line and it is cut-off (see arrow in Figure 5F).

The dispersion curve and propagation length of the same geometry, but with the silver modelled by the experimental permittivity is drawn in Figure 7E and F. As in the case of the asymmetric IMI structure, the actual one presented three modes. One of the modes that has no PSP character (solution 3). Other mode presents small wavevectors and a cut-off, but longer propagation lengths. Finally, a third mode with great wavevectors and PSP, that bends in the quasi-bound mode, reaching zero. This mode does not appear in the bulk region.

9. Conclusion

We showed, mathematically, the principal properties of propagating surface plasmons. We presented the analysis of plasmons that appear at planar interfaces, such as IM, symmetric and assymetric variants of IMI and MIM, and IIMI. Only for the IM geometry, there is an explicit function $\beta = f(\omega)$ (Equation 28). For the other geometries, the dispersions are implicit functions[7] (Equations 46, 52 and 64). For these geometries, one can re-arrange the dispersion relations as $f(\beta, \omega) = 0$, and numerically find the zeros for β in the first quadrant, as we can define the propagation towards $+x$, and lossy materials lead to exponential decays of the fields [32].

When modelling the metals as lossless by using the lossless Drude's permittivities, the propagation constants are real. In this case, the wavevectors parallel to the interface diverge at the surface plasmon frequencies. When the metals are modelled as lossy by using complex experimental values of permittivities, the wavevectors do not diverge any more, and the dispersion curves bend forming quasi-bond modes, regions with anomalous dispersion ($d\omega/dk < 0$).

Figure 8B shows an example the surface $|f(\beta, \omega)|$ for the IIMI structure at one specific frequency and using tabulated permittivity of silver. This surface presents many minima. The minima that converge to zero are all solutions of the dispersion relation. Infinite solutions are possible for one single frequency, but the most interesting ones are the solutions with relatively long propagation lengths — from tens of nanometers up. Nevertheless, the solutions with very short propagation lengths are also important for matching the boundary conditions and launching the propagating surface plasmons [32].

Author details

Baltar Henrique T. M. C. M., Drozdowicz-Tomsia Krystyna and Goldys Ewa M.
Faculty of Science of Macquarie University, Department of Physics and Astronomy, Sydney, Australia

10. References

[1] Aroca, R. F., Ross, D. J. & Domingo, C. [2004]. Surface-enhanced infrared spectroscopy, *Applied Spectroscopy* 58(11): 324A–338A.

[2] Bozhevolnyi, S. I. [2009]. *Introduction to surface plasmon-polariton waveguides*, Pan Stanford, chapter 1.

[3] Cheng, D. K. [1989]. *Field and wave electromagnetics*, 2 edn, Addison-Wesley.

[4] Dionne, J. A., Sweatlock, L. A., Atwater, H. A. & Polman, A. [2005]. Planar metal plasmon waveguides: frequency-dependent dispersion, propagation, localization, and loss beyond the free electron model, *Physical Review B* 72(7).

[5] Dionne, J. A., Verhagen, E., Polman, A. & Atwater, H. A. [2008]. Are negative index materials achievable with surface plasmon waveguides? a case study of three plasmonic geometries, *Optics Express* 16(23): 19001–19017.

[6] Fort, E. & Gresillon, S. [2008]. Surface enhanced fluorescence, *Journal of Physics D: Applied Physics* 41(1): 013001.

[7] Homola, J. [2006]. *Electromagnetic theory of surface plasmons*, Springer series on chemical sensors and biosensors, Springer, chapter 1.

[8] Huber, A., Ocelic, N., Kazantsev, D. & Hillenbrand, R. [2005]. Near-field imaging of mid-infrared surface phonon polariton propagation, *Applied Physics Letters* 87(8): 081103.

[9] Hwang, E. H. & Das Sarma, S. [2007]. Dielectric function, screening, and plasmons in two-dimensional graphene, *Physical Review B* 75(20): 205418.

[10] Iskander, M. F. [1992]. *Electromagnetic fields and waves*, Prentice-Hall.

---

[7] Functions not with the strict meaning, as one input may produce many outputs.

[11] Jablan, M., Buljan, H. & Soljačić, M. [2009].   Plasmonics in graphene at infrared frequencies, *Physical Review B* 80(24): 245435.

[12] Jackson, J. D. [1962]. *Classical electrodynamics*, 1 edn, John Wiley and Sons.

[13] Johnson, P. B. & Christy, R. W. [1972]. Optical constants of the noble metals, *Physical Review B* 6(12): 4370–4379.

[14] Junxi, Z., Lide, Z. & Wei, X. [2012]. Surface plasmon polaritons: physics and applications, *Journal of Physics D: Applied Physics* 45(11): 113001.

[15] Kawata, S. [2001]. *Near-field optics and surface plasmon polaritons*, Springer.

[16] Lavoie, B. R., Leung, P. M. & Sander, B. C. [2011]. Metamaterial waveguides, *ArXiv e-prints* .

[17] Le Ru, E. C. & Etchegoin, P. G. [2008]. Surface-enhanced raman scattering (sers) and surface-enhanced fluorescence (sef) in the context of modified spontaneous emission, *ArXiv e-prints* .

[18] Le Ru, E. C. & Etchegoin, P. G. [2009]. *Principles of surface-enhanced Raman spectroscopy and related plasmonic effects*, Elsevier.

[19] Liu, H., Genov, D. A., Wu, D. M., Liu, Y. M., Steele, J. M., Sun, C., Zhu, S. N. & Zhang, X. [2006]. Magnetic plasmon propagation along a chain of connected subwavelength resonators at infrared frequencies, *Physical Review Letters* 97(24): 243902.

[20] Lynch, D. W. & Hunter, W. R. [1985]. *Comments on the optical constants of metals and an introduction to the data for several metals*, Academic Press, Orlando.

[21] Maier, S. A. [2007]. *Plasmonics: fundamentals and applications*, Springer.

[22] Mills, D. L. & Burstein, E. [1974]. Polaritons: the electromagnetic modes of media, *Reports on Progress in Physics* 37(7): 817–926.

[23] Novotny, L. & Hecht, B. [2006].   *Principles of nano-optics*, Cambridge University, Cambridge.

[24] Pendry, J. B., Holden, A. J., Robbins, D. J. & Stewart, W. J. [1999]. Magnetism from conductors and enhanced nonlinear phenomena, *IEEE Transactions on Microwave Theory and Techniques* 47(11): 2075–2084.

[25] Rana, F. [2008].   Graphene terahertz plasmon oscillators, *IEEE Transactions on Nanotechnology* 7(1): 91–99.

[26] Rivas, J. G., Kuttge, M., Bolivar, P. H., Kurz, H. & Sánchez-Gil, J. A. [2004]. Propagation of surface plasmon polaritons on semiconductor gratings, *Physical Review Letters* 93(25): 256804.

[27] Sarid, D. & Challener, W. A. [2010]. *Modern introduction to surface plasmons: theory, mathematica modeling, and applications*, Cambridge.

[28] Sernelius, B. E. [2001]. *Surface modes in physics*, Wiley-VCH.

[29] Staelin, D. H., Morgenthaler, A. W. & Kong, J. A. [1994]. *Electromagnetic waves*, Prentice-Hall.

[30] van Vugt, L. K. [2007]. *Optical properties of semiconducting nanowires*, PhD thesis.

[31] Welford, K. [1991]. Surface plasmon-polaritons and their uses, *Optical and Quantum Electronics* 23(1): 1–27.

[32] Zakharian, A. R., Moloney, J. V. & Mansuripur, M. [2007]. Surface plasmon polaritons on metallic surfaces, *IEEE Transactions on Magnetics* 43(2): 845–850.

[33] Zayats, A. V. & Richards, D. [2008]. *Near-optics and near-field optical microscopy*, Artech House.

[34] Zayats, A. V., Smolyaninov, I. I. & Maradudin, A. A. [2005]. Nano-optics of surface plasmon polaritons, *Physics Reports* 408(3-4): 131–314.

Chapter 8

Light Transmission via Subwavelength Apertures in Metallic Thin Films

V. A. G. Rivera, F. A. Ferri, O. B. Silva, F. W. A. Sobreira and E. Marega Jr.

Additional information is available at the end of the chapter

1. Introduction

The optical properties of subwavelength apertures in metallic films have been the focus of much research activity around the world since the extraordinary optical transmission (EOT) phenomenon was reported over a decade ago (Ebbesen et al., 1998).

EOT is an optical phenomenon in which a structure containing subwavelength apertures in an opaque screen transmits more light than might naively be expected on the basis of either ray optics or even knowledge of the transmission through individual apertures. The phenomenon was discovered serendipitously for two-dimensional (2D) periodic arrays of subwavelength holes in metals (Garcia-Vidal et al., 2010). Surprisingly, such arrays may, for certain wavelengths, exhibit transmission efficiencies normalized to the total area of the holes that exceed unity. In other words, for these wavelengths a periodic array of subwavelength holes transmits more light than a large macroscopic hole with the same area as the sum of all the small holes. The surprise is compounded by the fact that a single subwavelength aperture generally transmits light with an efficiency that is substantially below unity.

This remarkable transmission enhancement has potential applications in photolithography, near-field microscopy, and photonic devices (Lal et al., 2007). Although the detailed picture of the transmission enhancement is still being investigated, the excitation of surface plasmon-polaritons (SPPs) is proposed to be involved in the process (Weiner, 2009). Ever since the first experimental report on EOT through subwavelength apertures, considerable theoretical effort has been devoted to interpreting the essential physics of the process in slit arrays (Porto et al., 1999; Takahura, 2001; Xie et al., 2005). Experimental studies subsequent to the initial report were also performed, which demonstrated a number of surprising features. For instance, spectral transmission measurements (Lezec et al., 2004) revealed that suppression, as well as enhancement, was a characteristic property of slit arrays. Additionally, interferometric studies (Gay et al., 2006a, 2006b; Kalkum et al., 2007) showed that the contribution of transient diffracted surface modes is as important as the SPP guided mode in the immediate vicinity of the subwavelength object. A more recent investigation (Pacifici et al., 2008) with the aim to confront the question of how transmission minima and maxima depend on array periodicity showed a minimum in transmission at slit separations equal to the wavelength of the SPP mode, and maxima occurring approximately at half-integer multiples.

These previous studies focused attention on the properties of arrays fabricated in metallic films. It was already shown that a single slit is an interesting structure, since it combines the compactness of a single defect with the directionality of an array launcher. The extended dimension of the long axis imposes directionality on the transmitted light beam, the divergence of which can easily be controlled (Laluet et al., 2008). In a very recent investigation dealing with single subwavelength slits, it was possible to observe that the slit transmission is notably affected by the film thickness, and increases linearly with increasing slit width for a fixed film thickness (Ferri et al., 2011).

In this chapter, we selected some fundamental subjects of high and general interest involved in the phenomenon of light transmission via subwavelength apertures in metallic thin films. The manuscript will cover issues on both theory and experiment, such as (1) fabrication and measurement setup characteristics, (2) different materials and metallic structures, and (3) phenomena from the metallic film/dielectric interface. In addition, numerical simulations were performed in order to investigate the optical transmission through subwavelength apertures.

The paper is organized as follows. In Section 2, we comment on the classical theories of diffraction by subwavelength apertures. In Section 3, a rapid discussion about preceding experiments and interpretations are presented. In Section 4, we give a simple introduction to the plasmon-polariton. In Section 5, some characteristics of focused-ion beam (FIB) nanofabrication are presented. In this manner, the topics covered by the above mentioned sections are basically a compilation of information found in literature. In Section 6, we in fact demonstrate some applications made by our research group investigating the (1) influence of the metallic film thickness, as well as the (2) use of multilayered metallic thin films on the optical transmission through subwavelength slits. A short summary is provided in Section 7.

2. Diffraction by subwavelength apertures

The wave nature of light implies modifications in the transmission through apertures like the phenomenon of diffraction. This common process, that even in simple geometries is very complex, have been extensively studied and many different models and approximations were developed based on the classical theory of diffraction (Jackson, 1999; Bouwkamp, 1954). Probably the simplest geometry, and maybe for this reason one that received the most attention, is that of a circular aperture with radius r in an infinitely thin and perfect conducting screen (Fig. 1).

If the radius r of the aperture is some orders of magnitude larger than the wavelength λ_0 of the impinging radiation, i.e., $r \gg \lambda_0$, the problem can be treated with the Huygens-Fresnel principle and its mathematical formulation can be given as a good approaching by the

Kirchhoff scalar theory of light diffraction (Jackson, 1999). This theory is based on a scalar wave theory, and, thus, it does not take into account effects due to the polarization of light. In the case of normal incidence of light through a circular aperture, it is easy to show that the transmitted intensity (dI) per unit solid angle (dΩ) in the far-field region, also known as Fraunhofer diffraction limit, is given by $I(\theta) \cong I_0 \left(k^2 r^2 / 4\pi \right) \left| 2J_1 (krsin\theta) / krsin\theta \right|^2$, where the incident intensity I_0 is equally distributed on the aperture area πr^2. The wavenumber of the incident light is k=$2\pi/\lambda_0$, θ is the angle between the normal to the aperture and the direction of the emitted radiation, and J_1(krsinθ) is the Bessel function of the first kind. The pattern described by this formula is the well known Airy pattern, composed by a central bright spot surrounded by concentric bright rings of decreasing intensity, caused by the interference of light rays originated inside the aperture (see Fig. 1). The ratio of the total transmitted intensity to I_0, given by $T = \int I(\theta) d\Omega / I_0$, is called the transmission coeficient. For large apertures, with $r \gg \lambda_0$, in which case the treatment is outlined here is valid, $T \approx 1$.

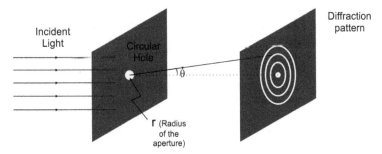

Figure 1. Transmission of light through a circular aperture or radius r in an infinitely thin opaque screen

It has been proposed (Weiner, 2009) that the existence of surface waves such as SPPs are involved in the transmission process. For this reason, the regime of subwavelength apertures r $\ll \lambda_0$ is much more interesting, because near-field effects are expected to contribute dominantly in the transmission process. The problem arising here is that even in an approximate analysis of a perfectly conducting screen in the limit of zero width, we must use a vector description via Maxwell's equations. In Kirchhoff's method, the basic assumption is that the electromagnetic field in the aperture is the same as if the opaque screen is not present, a case which does not fulfil the boundary condition of zero tangential electric filed on the screen. For large holes, in which $r \gg \lambda_0$, this basic failure is less severe, because the diffracted fields are small when compared to the directly transmitted ones. Nevertheless, for subwavelength apertures this approximation is inadequate even in a first order treatment of the problem.

Assuming that the incident light intensity I_0 is constant over the area of the aperture, Bethe and Bouwkamp arrived at an exact analytical solution for light transmission through a sub-wavelength circular hole in an ideal perfectly conducting and infinitely thin screen (Bethe, 1944; Bouwkamp, 1950a; Bouwkamp, 1950b). For normal incidence, the aperture can be described as a magnetic dipole located in the plane of the hole. The transmission coefficient for an incident plane wave is then given by (Maier, 2007)

$$T = \frac{64}{27\pi^2}(kr)^4 \propto \left(\frac{r}{\lambda_0}\right)^4 \qquad (1)$$

There is a weak total transmission for a subwavelength aperture due to the scaling with $(r/\lambda_0)^4$, smaller by an amount of $(r/\lambda_0)^2$ compared to Kirchhoff's scalar theory. This scaling of $T \propto \lambda_0^{-4}$ is in agreement with the theory of light scattering by small objects due to Rayleigh. The case described in Eq. (1) is that of normal light incidence with both transverse electric (TE) and transverse magnetic (TM) polarization. This is not the case when radiation is incident in the screen at another angle, in this case an additional electric dipole in the normal direction is needed to describe the different behaviour of the process. More radiation is transmitted for TM than for TE polarization in this case (Bethe, 1944).

There are two major approximations in the Bethe-Bouwkamp theory of light transmission through a circular aperture in a screen. The screen is said to be made of an ideal perfectly conducting screen, and so perfectly opaque to the transmission of radiation, and its thickness is taken to be infinitely small. One of these assumptions could be omitted by taking numerical simulations for the problem of screens with finite thickness (Maier, 2007). However, when discussing the transmission properties through real apertures, i.e., in real metals, the finite conductivity, and so the transmission, should be taken into account. The thin films used in optical experiments cannot be taken as perfectly opaque screens, and we could not employ the Bethe-Bouwkamp theory. On the other hand, if the film thickness is higher than some skin depths, that is, if we are dealing with a "thick" film, it could be taken as an opaque screen. It has been shown that for apertures fulfilling these conditions, localized surface plasmons have a significant influence in the transmission process (Degiron et al., 2004).

3. Early experiments and implications

In the paper published by Ebbesen et al., it was affirmed the measurement of "transmission efficiency...orders of magnitude greater than predicted by standard aperture theory". A typical result obtained in the original experiments is shown in the right panel of Fig. 2. The figure shows the transmitted power as a function of the incident light wavelength λ through a circular aperture of radius r. The transmission intensity is normalized to the cross-sectional area of the hole $A=\pi r^2$. Using the experimental parameters of the caption of Fig. 2, the efficiency predicted by Bethe theory, given by Eq. (1), is 0.34%, while the results obtained by Ebbesen et al. reported peak efficiencies of a factor more than two. In this case, the enhancement of light transmission over that expected by Bethe theory is about 600 times. In a following paper (Ghaemi et al., 1998) the same group reported peak transmission efficiencies "that are about 1000 times higher than that expected for subwavelength holes". To obtain an EOT it is plausible to say that would be necessary at least a three-order of magnitude increase over the predictions of Bethe formula.

It was proposed in these early reports that transmission enhancements were caused by a new phenomenon not taken into account in electromagnetic (E-M) vector field Bethe theory nor in the scalar diffraction Kirchhoff model: the resonant excitation of surface plasmons

(SP) waves supported by the periodic array structure in the metal film. The physics of SP waves shall be discussed in the next Section. About a year later Treacy suggested (Treacy, 1999) "dynamic diffraction" as another model to investigate the problem of light transmission through a periodic array of holes or slits. It was pointed out that the oscillation at frequency ω of the optical field would induce currents within the skin depth of the metal. The periodic structuring of holes in the metal gives rise to Bloch modes of the E-M field induced in the metal within the skin depth and consistent with its periodicity. There will be an oscillating current associated to each of these Bloch modes. Then, Treacy invoked "interband scattering" as a way to distribute these energy among these Bloch modes and from there to the propagating modes and surface waves at the aperture exit. It was not exactly clarified how does this redistribution occurs, but Treacy suspected that the success of "dynamic diffraction" for interpreting X-ray scattering in crystals might be useful for the understanding of light transmission through these new structures as well.

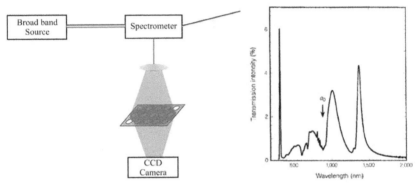

Figure 2. Left panel: schematic of the original spectral transmission experiments. A broadband, incoherent light source is spectrally filtered by a scanning spectrophotometer and focused onto an array of subwavelength structures (holes or slits). A charge-coupled device (CCD) camera records the transmission intensity through the structure as a function of wavelength of the input light (adapted from Weiner, 2009). Right panel: transmission spectrum from an array of holes in an Ag metal film evaporated onto a transparent substrate as reported by Ebbesen et al. Array periodicity: 900 nm, hole diameter: 150 nm, metal film thickness: 200 nm. The point indicated a_0 marks the array periodicity

A much more complete presentation of dynamic diffraction was published two years later by Treacy (Treacy, 2002). In this work it was clarified the relation between this approach and earlier interpretations of transmission in terms of "resonant" excitation of SP waves. This pioneering paper pointed out the way forward by emphasizing critical factors in the proper analysis of the problem. The most important thing to observe is that Bloch modes are determined by the E-M field present on and below the metal surface where are placed the periodic structures. These modes obey the periodic boundary conditions of the structure independently of the wavelength of the incident light. The simplest case is that of a one dimensional (1D) structure, with arrays and/or grooves placed periodically. In this case the modes are parallel to the surface in the metal and in the dielectric media, and evanescently vanishing in the normal direction. In parallel to this there are propagating modes. These propagating modes are responsible for the transmission of light through the structures. The transmission of the E-M filed is defined by both, propagating and evanescent fields, as a result of linear combinations of Bloch modes.

4. Plasmon-polaritons

It is important to stress here that metals play a more important role in plasmonics than dielectric media (Huang et al., 2007). In a metal, the optical as well as the electric properties are very different from dielectrics because of the existence of huge free electrons. These electrons have a fast response to varying fields leading to a different response than that in a dielectric media.

In the Drude model for free electrons, the dielectric constant is given by

$$\varepsilon_m = 1 - \frac{\omega_p^2}{\omega(\omega + i\gamma)} \qquad (2)$$

The constant ω_P is called the bulk plasma frequency and is a constant that depends on the metal. The constant γ is associated to the scattering of electrons in the Drude model. In "good" metals, where the scattering process is reduced we can neglect the damping ($\gamma = 0$). For high frequencies the dielectric constant is positive and there are modes whose dispersion relation is given by $\omega^2 = \omega_p^2 + c^2 k^2$. In this expression, c is the light velocity in vacuum. These modes are known as bulk plasmon-polaritons (BPPs) and are a result of the coupling between light and the free electrons in metals. For light at low frequencies, which is the case of visible light for metals, then $\omega < \omega_p$, and light propagation is forbidden by the negative permittivity.

For the case of a metal, whose real part of permittivity is negative, light incident normally to the surface gives rise to evanescent modes. But even in metals, there are propagating modes in the surface of the metal, provided that the surface of the metal is interfaced with a dielectric (or vacuum). These modes are the so-called SPP waves mentioned before. A typical geometry for this kind of problem is a metal and a dielectric separated by an infinite plane surface, as shown in the left panel of Fig. 3. This surface wave is based on the coupling between the surface free charges along the metal and light. In this case, the dispersion relation is given by (Zayats et al., 2005):

$$k_{SPP} = \frac{\omega}{c} \sqrt{\frac{\varepsilon_m \varepsilon_d}{\varepsilon_m + \varepsilon_d}} \qquad (3)$$

Here, ε_d is the permittivity of dielectric. The condition for the propagation of the SPP is that k_{SPP} is real. As the permittivity of metal is negative it is necessary that $\varepsilon_m + \varepsilon_d < 0$, in this case $\omega < \omega_{SP} \equiv \omega_p / \sqrt{1 + \varepsilon_d}$, where ω_{SP} is the surface plasmon frequency. The dispersion relation for the SPPs and BPPs are plotted schematically in the right panel of Fig. 3. The left panel of Fig. 3 shows the mechanism of how the surface plasmon propagates along the metal/dielectric surface.

Some features of the SPP propagation along these flat interfaces between a metal and a dielectric, as summarized by Huang et al., are addressed in the following.

It was discussed that SPP modes are the result of a coupling between the free electrons in the metal side and light (E-M field). The modes given by the dispersion relation in Eq. (3) are called TM modes. These are the only allowed modes, in the case of an interface between two non-magnetic media, and are characterized by a magnetic field normal to the propagation direction (in the left panel of Fig. 3 it is normal to the plane of the paper) and an electric filed that has components parallel ($E_{//}$) and normal ($E_\perp$) to the direction of propagation (the plane of the paper in Fig. 3). The ratio of the normal and parallel components of electric field inside the dielectric $(E_\perp/E_{//})_d = \sqrt{\varepsilon_m/\varepsilon_d}$ and inside the metal $(E_\perp/E_{//})_m = -\sqrt{\varepsilon_d/\varepsilon_m}$ shows that, inside the metal, as the electric field is almost completely concentrated in the direction of propagation, the free electrons present a movement of back and forth in the direction of propagation. These are the electron density waves shown in Fig. 3.

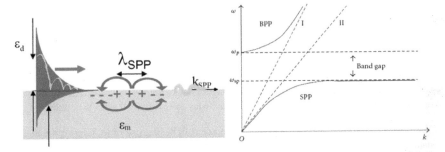

Figure 3. Plasmon-polariton modes associated with the metals. Left panel: SPP mode in a schematic view of infinite metal/dielectric interface. Right panel: dispersion relation of BPP and SPP modes. The dashed lines I and II denote the light dispersion $\omega = ck$ and $\omega = ck/\sqrt{\varepsilon_d}$, respectively. The frequency range between ω_{SP} and ω_P corresponds to a gap where the electromagnetic wave cannot propagate via either bulk or surface modes (Huang et al., 2007)

The SPP modes propagate along the interface between media with a larger propagation constant ($k_{SPP} > k_0\sqrt{\varepsilon_d}$), so the wavelength and propagation velocity are smaller than in vacuum. Taking into account that the metal permittivity still has a negative part, that is responsible for losses, the propagation of SPP is reduced to a finite value given by the propagation length $L_{SPP} \approx \varepsilon_m'^2/k_0\varepsilon_m''\varepsilon_d^{3/2}$, where ε_m' and ε_m'' are the real and imaginary parts of metal permittivity, respectively. In visible and near-infrared region, L_{SPP} can take values of several micrometers, as can be seen in Fig. 4, that shows the real and imaginary parts of permittivity for Au and Ag [Fig. 4(a)], and the propagation length for each of these media with an interface with air [Fig. 4(b)].

The confinement of SPP modes along the surface is characterized by its evanescent behaviour in either side of the interface, which happens because of the larger propagation constant. On the dielectric side there is a higher penetration of the E-M field, given by the decaying length $\delta_d \approx \sqrt{\left|\epsilon'_m\right|}/k_0\epsilon_d$. On the metal side, where the decaying length is $\delta_m \approx 1/k_0\sqrt{\left|\epsilon'_m\right|}$, there is a higher confinement than in the dielectric side. The relation between these lengths δ_d/δ_m is about some tens. This is what is desired in practice. Moreover, there is a strong enhancement of the fields near the interface of the media.

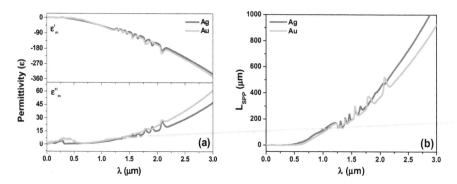

Figure 4. (a) Real ϵ'_m and imaginary ϵ''_m parts of permittivity for Au and Ag as obtained using data from Palik, and the corresponding (b) Propagation length LSPP as a function of the wavelength. In this case it is considered an interface between metal and air, whose permittivity is taken to be unity

SPP modes cannot be excited directly by the incident light because of its larger propagation constant. Special techniques have been used (Zayats et al., 2005) to compensate the phase mismatch as well as the difference in the wavevectors. Some of these techniques employ, for example, prism coupling by attenuated total reflection and diffraction gratings. It is also possible to couple plasmons by near-field excitation with a near-field optical microscope. An efficient mode to excite SPP is a subwavelength hole or slit (Zayats et al., 2005; Yin et al., 2004; Lalanne et al., 2005), where the diffraction components can ensure momentum conservation. Because of its localized feature, a direct observation of SPP is very difficult. Some methods, like the observation by using near-field microscopy (Hecht et al., 1996) can be used for this purpose. It is also possible to map SPP modes by recording the scattered light from a metal surface (Bouhelier et al., 2001) prepared with structured nanoscale corrugations. Another method, called fluorescence imaging (Ditlbacher et al., 2002a), has been proposed and consists in a metal surface covered with fluorescent molecules which emit radiation with intensity proportional to the electric field. This method was successfully used to observe interference, beam splitting and reflection of SPP modes (Ditlbacher et al., 2002b).

It is also important to note that even in more complex cases of metal/dielectric interfaces, SPP modes can exist with some parallel characteristics. A good example is that of a thin metal film where the thickness is of the same order of the skin depth. In this case, the SPP modes excited in each of its surfaces couple together and give rise to the so-called long-range SPP modes (Sarid, 1981). There is also a current research interest in particle plasmon-polariton modes that are excited in particles with have dimensions much smaller than the E-M wavelength. A detailed discussion on this topic can be found in (Hutter et al., 2004).

5. Focused-ion beam nanofabrication

Plasmonic structures can be obtained by many standard techniques, some of them are: optical lithography, electron-beam lithography, focused-ion beam (FIB) lithography, atomic layer deposition, soft lithography and template stripping (Lindquist et al., 2012) Taking into account that in the present work FIB lithography was extensively used, we will briefly discuss here only this fabrication technique. A comprehensive discussion of the other mentioned techniques (as well as FIB lithography) is very well presented in the work by Lindquist et al.

FIB lithography has been extensively used for direct fabrication of metallic nanostructures, by making patterns on substrates (Melngailis, 1987; Orloff et al., 1993; Langford et al., 2007). It can also be used for the deposition of various metals by using ion-beam induced deposition (Tao et al., 1990), for doping semiconductors (Melngailis, 1987; Moberlychan et al. 2007), and for preparing transmission electron microscope (TEM) samples (Reyntjens et al., 2001; Mayer, 2007). These methods proved to be very useful to make tests of device designs and geometries, fix masks or electrical traces, or to produce high-resolution ion-beam images, and have been an essential tool for the development of the field of plasmonics.

FIBs impinging on a surface offer a very different form of nanopatterning compared with other conventional methods that use resist, exposure and development. In general, the accelerating potential of the ions is of tens of kilovolts, the current beams range from many orders of magnitude, from picoamps to several nanoamps. Depending on the column optics, ion source and beam current, the beam spot sizes can range from ~ 5 nm up to a micrometer. Numerous ion species can be used in the setup, such as Al, Au, B, Be, Cu, Ga, Ge, Fe, In, Li, P, Pb and Si, the most commonly used being the semiconductor dopants (Melngailis, 1987; Orloff et al., 1993). In particular, Ga is widely used due to its low melting temperature (30 °C), low volatility and low vapour pressure (Volkert et al., 2007).

The FIB system is based on a liquid-metal ion source which is used to produce a smaller and brighter ion beam (Volkert et al., 2007). A metal source is heated up, such that it flows down and wets a sharp tungsten needle. An extraction voltage applied between the metal source and an extraction aperture forces the liquid metal to be pulled into an extremely sharp "Taylor–Gilbert" cone (Volkert et al., 2007; Forbes et al., 1996). The balance between the electrostatic force produced by the extraction voltage and the surface tension forces in the liquid (Orloff et al., 1993), the liquid source can have a tip size on the order of several nanometers (Melngailis, 1987). Then, the ions are extracted from the tip of this cone by field emission. As opposed to e-beam imaging systems, the "lenses" used in FIB are electrostatic and not magnetic, this happens because the Lorentz force in heavy ions, as those used, is much smaller than in electrons with the same kinetic energy (Volkert et al., 2007). Some other similarities and differences in these two kind of systems is presented in Fig. 5(a). To make the patterns a fast beam blanker is used. In commercial systems many patterns can be drawn by using pre-fabricated CAD proprietary files or using options within the control software. The final size and the resolution of the focused beam can, in general, be affected by chromatic aberration, i.e., the energy dispersion of ions in the beam, but with a good approximation it can be regarded as a Gaussian profile. The minimum beam size is on the order of ~ 5 nm. A schematic view of the ion source is shown in Fig. 5(b).

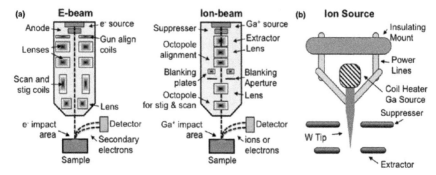

Figure 5. (a) Schematic view showing the similarities and differences between typical e-beam and FIB systems. (b) Scheme of the gallium liquid metal source (Lindquist et al., 2012)

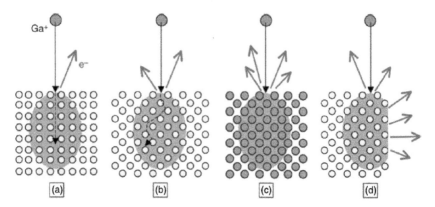

Figure 6. The crystal orientation of a sample can affect the FIB sputtering rates, shown in (a) and (b). (c) The sputtering rate is also affected by the mass of the atoms (orange atoms are more massive) and (d) by the local geometry of the sample (Lindquist et al., 2012). Original figure from Volkert et al.

In the process of milling a substrate by using FIB, many effects can be produced, with approximately one to five atoms removed per incident ion depending on the ion energy or substrate. It is even possible to displace atoms from their equilibrium positions, to induce chemical reactions and use the emerging electrons for imaging. The Gaussian profile of the beam is not the only factor that the shape of the milled groove depends (Melngailis, 1987). Redeposition and self-focusing effects can lead to large geometric differences depending on whether the patterns designed were done in single of multiple steps, even with the same overall dose. Also, milling a trench with large total ion doses deviates from the Gaussian profile, giving an unexpectedly deep, V-shaped groove (Melngailis, 1987). The FIB milling process depends on many aspects such as the material to be patterned, ion-beam incident angle, redeposition of sputtered materials and even the crystal orientation, as outlined in Fig. 6. Grain orientation-dependent FIB sputtering can also lead to severe surface roughness on polycrystalline samples, however, care can be taken to produce well-defined structures (Lindquist et al., 2012).

FIB can also be used for the deposition of various metals (such as W, Pt, C and Au) via site-specific chemical vapour deposition (CVD), this can be done by using a gas injection system (GIS) (Volkert et al., 2007). A high-efficiency deposition can be achieved, with about 1 μm per minute accumulation rates, by adjusting the gas precursor flow rates and the ion-beam current density. The reaction of the ion beams with the precursor materials offers the ability to weld micromanipulators to specific parts of a substrate in situ. With subsequent FIB milling and thinning, those parts are cut free, and are often mounted to a TEM imaging grid (Mayer, 2007). A deficiency of this kind of system is that metals deposited via FIB have a high contamination by carbon. It can also be used other GIS systems, instead of depositing metals, to enhance the inherent FIB milling etch rate.

FIB instruments offer many significant advantages, like a direct write, maskless, high-resolution nanofabrication with the ability to sputter, image, analyse and deposit. It is possible to design 2D and 3D patterns (Langford et al., 2007). However, it has some limitations as a patterning tool, particularly for metals. FIB milling is a serial lithography technique, such as e-beam lithography, patterning only one spot or device at a time, unlike optical lithography that patterns the whole wafer with one short exposure. Large area patterning is not feasible. For high-resolution (< 100 nm) features, FIB milling can also be slow since very low currents (~ pA) must be used. Along with FIB-induced sample damage (Mayer, 2007), Ga ions are implanted at atomic fractions of 1–50% near the sample surface (Volkert et al., 2007). The plasmonic properties of the patterned metal films can be degraded is this process. For high surface roughness of metals the SPP propagation length can be strongly reduced (Lindquist et al., 2012). As such, the advantages and disadvantages of FIB milling need to be taken into account when fabricating new optical or electronic devices. When combined with a template stripping technique, many of these roughness and contamination issues are minimized, since FIB is then used to only pattern a reusable template, leaving the resulting metal films smooth and contamination-free (Lindquist et al., 2012).

6. Applications: Optical transmission through subwavelength single slits

6.1. Influence of metallic film thickness

6.1.1. Motivation

By focusing attention on the properties of arrays fabricated in metallic films with fixed thickness, some previous studies mentioned in Section 1 apparently missed the important role played by the film thickness. For example, in the designing of surface-plasmon-based sensors, a proper choice of the thickness of the metallic film for the optimization of the device sensitivity is very important (Fontana, 2006). There are only a few theoretical (Xie et al., 2005; Fontana, 2006; Janssen et al., 2006) and experimental (Shou et al., 2005; Kim et al., 2006; Pang et al., 2007) investigations taking into account the influence of film thickness on the considered process. Therefore, the present studies are motivated by the necessity to understand the physics of this phenomenon and to develop optimum designs practices for subwavelength structure fabrication. We present here a systematic study of the optical transmission through subwavelength slits fabricated in Ag and Au film samples possessing different thicknesses. The influence of slit width was also considered. The present work deals with single slits. The extended dimension of the long axis imposes directionality on the transmitted light beam, the divergence of which can easily be controlled (Laluet et al., 2008). A slit is thus an interesting structure since it combines the compactness of a single defect with the directionality of an array launcher. Moreover, in order to remove measurement ambiguities existing in setups employing an incoherent and broadband light source dispersed through a scanning spectrophotometer, as shown in the left panel of Fig. 2, we have measured the transmission intensity through the slits using coherent and monochromatic spectral sources (Pacifici et al., 2008; Ferri et al., 2011).

6.1.2. Experimental considerations

A series of Ag films with thicknesses of 120, 160, 200, 270, and 330 nm and a set of Au samples with thicknesses of 120, 180, 260, 360, and 450 nm, as measured by a Talystep profilometer, were thermally evaporated onto BK7 glass substrates. Slits with widths in the range of approximately 70–150 nm in the Ag films, and 120–270 nm in the Au films, were milled with an FEI focused ion beam QUANTA 3D 200i (Ga+ ions, 30 keV). In order to verify the depth of the slits, the gallium ions' source was calibrated using atomic force microscopy. For example, the right panel of Fig. 7 shows a scanning electron micrograph of a slit with 150 nm of width fabricated in the 200 nm thick Ag film.

We have undertaken a series of high-resolution measurements of the optical transmission through the slits. The transmission measurement setup consists of 488.0 nm (for Ag) and 632.8 nm (for Au) wavelength light beams from Ar ion and HeNe lasers, respectively, with a power of about 1 µW, aligned to the optical axis of a microscope. The beam is focused at normal incidence onto the sample surface by a 20× microscope objective (with an NA of 0.4) in TM polarization (magnetic field component parallel to the long axis of the slits). Light intensity transmitted through each slit is then gathered by an optical fibre and detected with a CCD array detector. It was used on a multimode fibre with an NA of 0.22 and a core diameter of 200 µm. Light intensity is obtained by integrating the signal over the entire region of interest in the CCD image and subtracting the background originating from electronic noise. The transmitted intensity of every slit was recorded in the far-field by the CCD as the sample was stepped using an x-y translation stage. The left panel of Fig. 7 shows the schematic of the measurement setup.

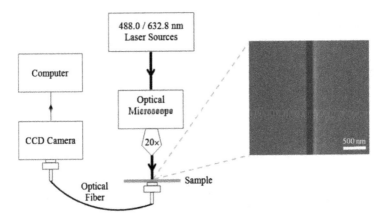

Figure 7. Left panel shows a schematic of the optical transmission experiment. 488.0 nm (for Ag) and 632.8nm (for Au) Ar ion and HeNe laser light sources, respectively, are normally focused onto the sample surface by a 20× microscope objective lens. A CCD camera records the transmission intensity through the slits as the sample surface was stepped. Right panel shows a scanning electron micrograph (taken with 40000× magnification) of a typical structure. The considered slit has approximately 150 nm of width and was focused-ion-beam milled through a 200 nm thick Ag layer. In the experiments, the thicknesses of the Ag and Au films were varied in the range of 100–450 nm. The width is varied from 70 to 270 nm (Ferri et al., 2011)

6.1.3. Results and discussion

Fig. 8 shows the physical picture adopted in this work to investigate the light transmission through the subwavelength slits. The essential elements of the model, which describes a plasmonic damped wave with amplitude decreasing as the inverse of the film thickness (Gay et al., 2006b), are represented in the sketch of Fig. 8. Basically, an incident monochromatic light beam with wave vector k_0 in air is linearly polarized perpendicular to the slit of subwavelength width w, milled in a metallic film with thickness t, and deposited on a dielectric substrate (BK7 glass).

The far-field intensity enhancement for the single slits involves multiple coupling processes (see Fig. 8). Initially, the incident laser light generates SPs on the metal film. Because of vertical plasmon coupling, which depends on the film depth (t), surface charges are induced on the top metal film and simultaneously a strong electric field is generated inside the slit. Subsequently, an SPP mode (Maier, 2007), i.e., an electromagnetic excitation propagating at the interface between the dielectric and the metallic conductor, evanescently confined in the perpendicular direction, is generated on the metal film/BK7 interface. The SPP evanescent mode travels along the interface toward the slit, where it reconverts to a propagating wave and interferes with the travelling field directly transmitted through the slit. Additionally, penetration of the incident field inside the film enables the excitation of localized SP resonances (Maier, 2007) on the rim of the aperture, which contribute to the superposed output field. In this way, induced dipole moments at each rim form an "antenna coupling", which radiatively generate strong field enhancement (top and bottom). Then, the intensity of the resulting field can be written as

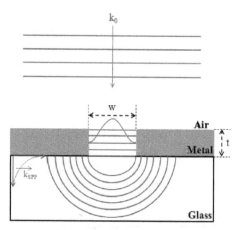

Figure 8. Illustration of the adopted model. A single frequency incoming plane wave with wave vector k_0 in air is linearly polarized perpendicular to a slit of subwavelength width w, milled in a metallic film with thickness t deposited on a BK7 glass substrate. Here, k_{SPP} is the wavevector of the SPP mode. (Ferri et al., 2011)

$$ E \approx \frac{E_0}{\pi} \frac{w}{t} \cos\left(k_{SPP} t + \frac{\pi}{2} \right) \tag{4} $$

where $k_{SPP} = 2\pi/\lambda_{SPP}$, $\lambda_{SPP} = 2\pi/\mathrm{Re}\left[\beta\right]$, $\beta = k_0 n_{SPP}$, $n_{SPP} = \left(\varepsilon_{metal}\varepsilon_{glass}/\varepsilon_{metal} + \varepsilon_{glass} \right)^{1/2}$, and $k_0 = 2\pi/\lambda_0$ (Gay et al., 2006b; Pacifici et al., 2008; Maier, 2007). Here, E_0 represents the electrical field of the incoming plane wave, where λ_0 is its wavelength. Also, k_{SPP} and λ_{SPP} are the wavevector and wavelength of the SPP, β is the propagation constant of the superposed travelling wave, and n_{SPP} is the effective index of the SPP, which is for the interface between the metal and dielectric. In addition, ε_{metal} and ε_{glass} are the dielectric permittivities of metal and glass, respectively, and are functions of the excitation wavelength. In this sense, $\varepsilon_{Ag} = -7.89 + 0.74i$ and $\varepsilon_{glass} = 2.31$ are the tabulated dielectric constants of Ag and BK7 glass in the wavelength of 488.0 nm. In the same way, $\varepsilon_{Au} = -9.49 + 1.23i$ and $\varepsilon_{glass} = 2.29$ are the corresponding dielectric constants of Au and BK7 in 632.8 nm (Palik, 1985). Here it is important to point out that the wavelength values of 488.0 and 632.8 nm are known to be close to the plasmon excitation wavelength of Ag and Au, respectively. Increasing the metallic film thickness leads to decoupling of the top and bottom antenna.

For illustration purposes, Fig. 9 shows corresponding 2D numerical simulations carried out with Comsol Multiphysics® for TM-polarized waves for a 150 nm slit fabricated in an Ag film when illuminated by the line at 488.0 nm of an Ar ion laser. Fig. 9(a) shows the amplitude of the magnetic H field (along the z direction). Figs. 9(b) and 9(c) shows the amplitude of the electric E field (in the y direction), with its vector representation in the x-y plane. Fig. 9(a) shows how the incident plane wave is modified by the existing subwavelength slit. It is possible to see that the considered wave is almost completely reflected from the unstructured part of the film. Around the slit entrance, the amplitude of the standing wave is markedly attenuated, where some lightwave transmission to the exit facet is apparent. On the dielectric/metal interface, a train of surface waves (SPPs) is evident together with waves propagating into space. In the rims, different charge configurations can be obtained, which can be symmetric or antisymmetrically coupled (Prodan et al., 2003). This coupling leads to determined charge configurations in each rim of the slit (top and bottom). From Figs. 9(b) and 9(c), these surface modes are clearly seen. It is possible to notice from the figures these resonances [antisymmetrically and symmetrically coupled in Figs. 9(b) and 9(c), respectively] on the facets of the slit. These modes are associated with localized SPs, which are nonpropagating excitations due to direct light illumination of the conduction electrons of the metallic nanostructure coupled to the electromagnetic field (Maier, 2007). A similar behaviour was observed in the simulations for Au films when

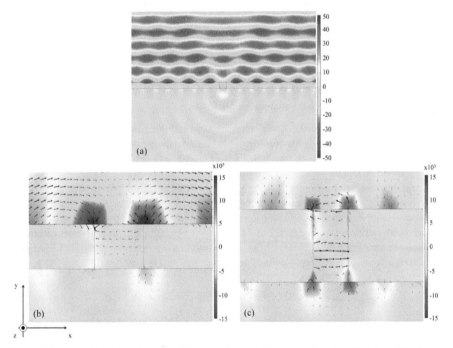

Figure 9. 2D numerical simulations of a 150 nm slit fabricated in an Ag film when illuminated by the line at 488.0 nm of an Ar ion laser. (a) Amplitude of the magnetic H field (along the z direction). (b) and (c) Amplitude of the electric E field (in the y direction), and its vector representation in the x-y plane. The value of the Ag film thickness in (a) and (b) is 120 nm, and in (c) is 270 nm. Length spans: (a) x = 4 μm and y = 2 μm, (b) and (c) x = 600 nm and y = 400 nm

excited by the line at 632.8 nm of an HeNe laser. The main apparent difference is the lower transmitted intensity due to a higher absorption loss attributable to the particular characteristics of Au (Palik, 1985), i.e., Re[ε_{Au}] = - 9.49 in contrast to Re[ε_{Ag}] = - 7.89. This is verified in Fig. 10, where it can be seen that the normalized transmission intensity for the Ag films is improved more than for the Au samples.

In general, the present numerical simulations qualitatively show appreciable light transmission through the slits. Actually, it was experimentally observed that the transmission sensitively depends on the metallic film thickness and slit width. As a first approximation, the theoretical slit transmission intensity can be given simply by the square modulus of Eq. (4). In this way, Fig. 10 plots as predicted [from Eq. (4)] and measured (using the setup shown in Fig. 7) transmission intensities as a function of the film thickness for the various slit widths milled in the Ag and Au samples. Also, the insets of Fig. 10 show the measured transmission versus slit width for certain film thicknesses. The relative slit transmission intensities are obtained by subtracting the background originating from the metal film and normalizing to the intensity from the wider slit structures. It is valuable to notice from Fig. 10 the very good correspondence between the theoretical estimate and the experiment. Therefore, taking into account the errors associated with the experimental determination of film thickness, slit width, and optical transmission intensities, it is possible to affirm that (1) the slits' transmission varies with metallic film thickness and presents a damped oscillatory behaviour as the film thickness increases, and (2) the transmission increases linearly with increasing slit width for a fixed metallic film thickness. Although the general behaviour is similar, distinct optical properties [see Fig. 4(a)] lead to perceptible differences in the transmitted intensity and the position of maxima and minima between the Ag and Au films.

To help in elucidating the first observation, it is valuable to note that Fabry–Perot (FP) resonances are expected to contribute to the enhanced transmission of subwavelength slit arrays (Porto et al., 1999; Takahura, 2001; Pang et al., 2007; Garcia-Vidal et al., 2002). In this way, FP modes related to the finite depth of the slits in the present films should give rise to transmission maxima at certain wavelengths (Garcia-Vidal et al., 2002). Actually, an accurate analysis recently published shows that the two maxima observed in Figs. 10(a) and 10(b) correspond to FP-like resonances within the slit volume for the first half-wave and full wave of the light within the slit (Weiner, 2011). In this sense, the FP multiple reflection effect within the slits leads to significant modulation of the transmission as a function of metal film thickness. These transmission maxima occur if the FP resonance condition is fulfilled (Li et al., 2009):

$$2k_0 Re\left[\frac{\lambda_0}{\lambda_{SPP}}\right] t + arg\left(\phi_1 \phi_2\right) = 2m_y \pi \tag{5}$$

where m_y (the FP mode) is an integer and ϕ_1 and ϕ_2 denote the phase of the reflection coefficients of the slit at the incident and output interfaces, respectively. Thus, the effect of slit depth on the transmission enhancement can be easily understood. When the incident wavelength and slit depth are satisfied by Eq. (5), a transmission maximum will occur. Furthermore, from Eq. (5), it is expected that the transmission under a certain incident wavelength has a period of $\lambda_{SPP}/2$ as a function of slit depth. The results of Fig. 10 can now be simply explained for the present Ag and Au films. Therefore, when the film thickness is near half- or full-integer wavelengths of the guided mode within the slit "cavity", optimal transmission is achieved, which implies a field enhancement inside the slit.

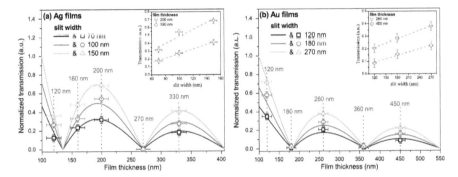

Figure 10. Theoretically estimated (lines) and experimental (symbols) normalized slit transmission intensities versus film thickness for the slits of the (a) 120, 160, 200, 270, and 330 nm thick Ag films, and the (b) 120, 180, 260, 360, and 450 nm thick Au samples. The dotted straight lines point out the thickness of the considered samples. The insets show the measured transmission versus slit width for some film thicknesses. Here, the dashed straight lines are linear fittings of the experimental points (Ferri et al., 2011)

We can apply an FP analysis to obtain the finesse F from Fig. 10, given by

$$F = \frac{\pi}{2\sin^{-1}\left(1/\sqrt{f}\right)} \tag{6}$$

where $f = 4R(1-R)^{-2}$ is the finesse factor. Here, R is the reflectivity, given by $R = 1 - T$, where T is the transmission (Born et al., 1993). We determine from Fig. 10(a) for the Ag samples with 100 nm of slit width and thicknesses of 120, 160, 200, 270, and 330 nm, the reflectivities $R = 0.73$, 0.67, 0.46, 0.99, and 0.73, and the corresponding finesses $F = 9.98, 7.74, 3.84, 312.58$, and 9.89, respectively. From Fig. 10(b) for the Au samples with 120nm of slit width and thicknesses of 120, 180, 260, 360, and 450 nm, we determine the reflectivities $R = 0.66, 0.99, 0.79, 0.98$, and 0.91, and the resultant finesses $F = 7.45, 312.58, 13.27, 155.49$, and 33.29, respectively. Here it is important to point out that for an FP cavity, the definition of quality factor (Q factor) is equivalent to the finesse (Shyu et al., 2011). Therefore, we can clearly see that both the R and Q factor values are significantly affected by the film thickness for a fixed slit width. Also, it is interesting to notice that near-zero transmission is a sign of high reflectivity values and high Q factors. For the maximum transmission points, a backward reasoning applies.

Finally, the fact that the transmission increases linearly with increasing slit width is in accordance with literature (Kihm et al., 2008), where it was observed that the far-field transmitted intensity from a single slit shows a monotonic increase with the width, as expected from macroscopic intuition. In other words, the physical cavity length L_z and the optical cavity length $L_c(\lambda)$ are related, such that $L_c(\lambda) = L_z + 2\delta(\lambda)|r(\lambda)|^2$, where $r(\lambda) = |r(\lambda)|\exp(i\phi)$ is the Fresnel coefficient, describes the shift of resonance wavelength from a perfect metal reflector due to field penetration $\delta(\lambda)$ into the metal mirror. But L_z is constant for all studied samples (20 μm), resulting in ϕ_z constant for a determined depth of metallic film. Nevertheless, the monotonic increase with the width w (= L_x) can also be explained considering FP resonances, for which we will use a simple analytical model to investigate the experimental results based on geometric arguments. Considering the standing wave mode in the cavity, when the penetration depth is ignored, the resonant condition of the slits can be written as

$$\frac{1}{\lambda_{SPP}^2} = \left(\frac{m_z + \phi_z}{2L_z}\right)^2 + \left(\frac{m_x + \phi_x}{2L_x}\right)^2 \qquad (7)$$

We have applied Eq. (7) using the data of 200 nm of depth with widths L_x = 70, 100, and 150 nm for an Ag film, and 260 nm of depth with widths L_x = 120, 180, and 270 nm for an Au film. L_z = 20 μm for both samples. Also, we considered in our calculations that ϕ_z and ϕ_x are practically constant, since we did not observe any peak shift in the transmission spectra of the samples in analysis. In the interface, values of λ_{SPP-Ag} = 269.9 nm and λ_{SPP-Au} = 361.6 nm were used. It was obtained for the Ag film m_z = 78 and m_x = 1, 1.4, and 2, and for the Au film m_z = 145 and m_x = 1, 1.5, and 2, i.e., an increase in L_x allows an increase in the transmission intensity spectra for a fixed depth [see insets of Figs. 10(a) and 10(b)].

6.2. Multilayered metallic thin films

6.2.1. Motivation

Although the detailed picture of the transmission enhancement is still being investigated, the excitation of SPPs on the two surfaces of the metal film has been proposed to be involved in the process (Moreno et al., 2004; Lezec et al., 2004). In fact, it was already shown that when two perforated metal films are spaced by a dielectric layer (cascaded metallic structure), the transmission is further increased compared to a single perforated metal film (Ye et al., 2005). Additionally, bimetallic structures, such as films and nanoparticles, have attracted considerable attention for plasmon resonance excitation (Zynio et al., 2002; Gupta et al., 2005; Tan et al., 2007; Chen et al., 2010). These works are focused on improving sensitivity and evanescent field enhancement by optimization of the thickness of Ag/Au layers. Recently, optimization in terms of spectral characteristics was also theoretically demonstrated (Dyankov et al., 2011).

Here, we propose novel structures providing a unique opportunity to generate plasmonic modes. The structures are based on subwavelength slits fabricated in multilayered metallic thin films. The main feature of the novel structure is that the metal film consists of alternating layers of Ag and Au. In this Section, we demonstrate that the slits transmission can be augmented by increasing their widths, with the advantage to offer minor losses in comparison with a single perforated metal film.

6.2.2. Experimental considerations

The experimental procedure adopted here is similar to that shown in Section 6.1.2. Multilayered Ag/Au/Ag/Au and Au/Ag/Au/Ag films with total thickness t = 200 nm (d = 50 nm for each layer) were thermally evaporated onto BK7 glass substrates. Slits with widths in the range of approximately 60–600 nm were milled in the films using FIB lithography with the same already mentioned conditions. The slit length was fixed at 5 μm. For example, the right panel of Fig. 11 shows a scanning electron micrograph of a slit with width w about 180 nm fabricated in the Ag/Au/Ag/Au film.

The transmission measurement setup is identical to that shown in the left panel of Fig. 7. Here, the 488.0 nm wavelength light beam from the Ar ion laser was used for the Ag/Au/Ag/Au film, and the 632.8 nm wavelength light beam from the HeNe laser was used for the Au/Ag/Au/Ag sample. The left panel of Fig. 11 shows the basic schematic of the experimental setup.

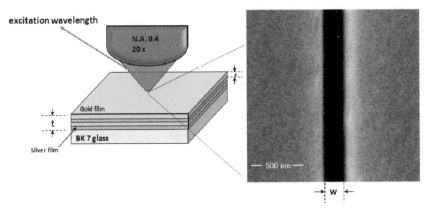

Figure 11. Left panel shows a simplified schematic of the optical transmission experiment. Similarly to the left panel of Fig. 7, a 488.0 nm (for the Ag/Au/Ag/Au film) and a 632.8 nm (for the Au/Ag/Au/Ag sample) Ar ion and HeNe laser light sources, respectively, are normally focused onto the sample surface by a 20× microscope objective lens. Right panel shows a scanning electron micrograph (taken with 60000× magnification) of a typical structure. The considered slit has approximately 180 nm of width and was focused-ion beam milled through the Ag/Au/Ag/Au sample. In the experiments, the total thickness t of the Ag/Au/Ag/Au and Au/Ag/Au/Ag films was fixed at 200 nm. The width w is varied from 60 nm to 600 nm (Ferri et al., 2012)

6.2.3. Results and discussion

Fig. 12(a) shows the physical picture adopted in this work to investigate the light transmission through the subwavelength slits fabricated in the multilayered metallic films. The essential elements of the model are represented (Gay et al., 2006b). Basically, an incident monochromatic light beam in air is linearly polarized perpendicular to the slit of subwavelength width w, milled in a multilayered metallic film with thickness t, deposited on a dielectric substrate (BK7 glass). Each metallic layer has a thickness d (= 50 nm).

The physical processes involved in the far-field intensity enhancement for the present subwavelength slits are similar to those discussed in Section 6.1.3, in the sense that the first metallic layer is responsible to start the generation of plasmonic excitations. Form our simulation results we can see that at the interface between the adjacent metallic layers the condition for generation of SPPs is not fulfilled, Fig. 12(b). Nevertheless, there is an induced charge current due to the plasmonic surface excitations from the first to the last metallic layer, see Fig. 12(c), resulting in an asymmetric distribution [similar to Fig. 9(b)]. Additionally, the transmission over each layer is given by the Beer-Lambert law, i.e., $I_i = I_{i-1}\exp(-\alpha d)$ (Born et al., 1993), where α is the absorption coefficient of the corresponding metal layer, d is the thickness of the layers, and i is the layer number. Then, the resulting transmitted intensity can be written as (Gay et al., 2006b; Maier, 2007; Pacifici et al., 2008; Ferri et al., 2011),

$$I_{out} = I_i + \left| \frac{E_0}{\pi} \frac{w}{t} \cos\left(k_{SPP}t + \frac{\pi}{2} \right) \right|^2, i = 1,...,4 \qquad (8)$$

Here, the physical quantities of the second term are identical with those in Eq. (4). The only difference is that the considered dielectric medium is the air, which tabulated dielectric constant is $\varepsilon_{air} = 1.00$ (Palik, 1985). In this sense, the first layer is assumed to govern the extraordinary transmission (I_{SPP}) of the subwavelength slits fabricated in the present multilayered metallic films.

Numerical simulations carried out with Comsol Multiphysics® were also performed for the multilayered metallic samples. A similar behaviour was observed in comparison to that shown in Section 6.1.3. For illustration purposes, Figs. 12(b) and 12(c) show simulations for a 50 nm slit fabricated in the Au/Ag/Au/Ag film when illuminated by the line at 632.8 nm typical of an HeNe laser. Fig. 12(b) shows the amplitude of the magnetic H field (along the z direction). Fig. 12(c) shows the amplitude of the electric E field (in the y direction), with its vector representation in the x-y plane.

The theoretical slit transmission intensity can be given by Eq. (8). In this way, Fig. 13 plots as predicted and measured transmission intensities as a function of the slit width for the various subwavelength structures milled in the Ag/Au/Ag/Au and Au/Ag/Au/Ag samples. For comparison purposes, the insets of Fig. 13 also show the simulated slit optical transmission obtained from Comsol Multiphysics® versus slit width for the considered multilayered metallic films and single perforated Ag and Au films with 200 nm of thickness. The relative slit transmission intensities are obtained by subtracting the background originating from the metal film and normalizing to the intensity from the wider slit structures. It is valuable to notice from Fig. 13 the very good correspondence between the theoretical estimative and experiment. Therefore, it is possible to affirm that: (1) the transmission increases linearly with increasing slit width, and (2) for a fixed width, the transmission of the multilayered structures is augmented in comparison with a single perforated metal film of the same thickness. It is evident that this last observation is more apparent for the Au/Ag/Au/Ag film.

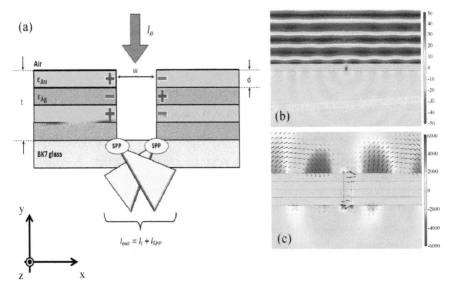

Figure 12. (a) Illustration of the adopted model. A single frequency incoming plane wave with intensity I_0 in air is linearly polarized perpendicular to a slit of subwavelength width w, milled in a multilayered metallic film with thickness t deposited on a BK7 glass substrate. Each metallic layer has a thickness d. The overall transmitted intensity I_{out} is a sum of the transmission of each layer I_i with the contribution due to the SPPs created in the air/metallic film interface I_{SPP}. (b) and (c) 2D simulations of a slit fabricated in the Au/Ag/Au/Ag film. (b) Amplitude of the magnetic H field (along the z direction). (c) Amplitude of the electric E field (in the y direction), and its vector representation in the x-y plane. Length spans: (b) x = 4 μm and y = 2 μm, and (c) x = 600 nm and y = 400 nm (adapted from Ferri et al., 2012)

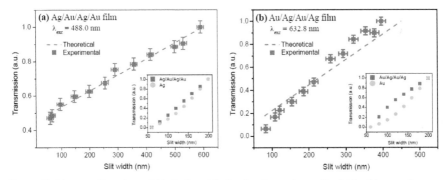

Figure 13. Theoretically estimated (dashed straight lines) and experimentally obtained (squares) normalized slit transmission intensities versus slit width for the various subwavelength structures milled in the (a) Ag/Au/Ag/Au and (b) Au/Ag/Au/Ag samples. The insets show details of the simulated slit optical transmission versus slit width for the considered metallic multilayered films and single perforated Ag and Au films with 200 nm of thickness (Ferri et al., 2012)

The fact that the far-field transmitted intensity from the present slits shows a monotonic increase with their widths is in the same trend of that previously observed in single perforated metal films (Ferri et al., 2011; Kihm et al., 2008), as expected from macroscopic intuition. In that case, the dependence could be explained considering Fabry-Perot resonances of the standing wave mode in the slit "cavity", in conjunction with the generation of SPPs. However, in the present case, such a resonant condition (Ferri et al., 2011; Kihm et al., 2008; Li et al., 2009) cannot be applied, since we have distinct reflection coefficients due to the existence of different materials in the slits. Nevertheless, the monotonic increase with the width is expected simply by considering the dependence of the transmitted intensity with the w parameter in Eq. (8).

Finally, the observation that the transmission of the metallic multilayered structures is augmented in comparison with a metal film of the same thickness when perforated with subwavelength slits, can be elucidated considering that, for the present multilayered films, the optical transmission profile is assumed to be mainly governed by the first metallic layer,

given that it is responsible to start the generation of plasmonic excitations. Subsequently, we just have electronic conduction to the underlying metallic layers, as already pointed out. Furthermore, each metallic layer (with 50 nm of thickness) additionally contribute to the overall transmission according to the Beer-Lambert law, in contrast to a single perforated metal film with the same total thickness of 200 nm of the proposed multilayered films.

Independently of the preceding discussion about multilayered metallic thin films, it is valuable to mention that gain-assisted propagation of SPPs at the interface between a metal and a dielectric with optical gain have been the focus of much research activity (Avrutsky, 2004; Nezhad et al., 2004). In this context, Er^{3+}-doped tellurite glasses as the dielectric medium is very attractive (Wang et al., 1994). Very recently, our research group have gave significant contributions regarding the excitation and/or improvement of the luminescence of Er^{3+} ions embedded in these glassy matrices through plasmonic nanostructures (Rivera et al., 2012a; Rivera et al., 2012b).

7. Conclusion

The physics of the transmission of light through subwavelength apertures in metallic films has been a topic of intense research in recent times. In this chapter, we have presented a review of this field, showing some essential subjects involved in this phenomenon. Although the current understanding of this phenomenon is not complete or even not substantially correct, the materials presented in many literatures are useful and clue us on how to go ahead. In particular, we presented in this chapter some contributions of our research group regarding the optical transmission through subwavelength single slits in metallic thin films. The simulations qualitatively reveal that the transmission profile is controlled by interference between the incident standing wave and plasmonic surface excitations. It was possible to observe that the slits' transmission is significantly affected by the metallic film thickness, presenting a damped oscillatory behavior as the film thickness is augmented. In addition, for a fixed metallic film thickness, the transmission increases linearly with increasing slit width. For a fixed wavelength and slit width, FP modes within the slits lead to significant modulation of the transmission as a function of metal film thickness. As well, it was shown that the transmission of multilayered structures is augmented in comparison with a single perforated metal film with a similar thickness. In this sense, we have demonstrated that metallic multilayered structures have the advantage to offer minor losses in comparison with a single perforated metal film.

Author details

V. A. G. Rivera, F. A. Ferri, O. B. Silva, F. W. A. Sobreira and E. Marega Jr.
Instituto de Física de São Carlos, Universidade de São Paulo, São Carlos, Brazil

Acknowledgement

The authors are indebted to Prof. J. Weiner, Prof. A. R. Zanatta and Dr. M. A. Pereira-da-Silva (all at the Instituto de Física de São Carlos, USP, Brazil) for the helpful discussions, optical transmission experiments and atomic force microscopy measurements, respectively. We also would like to thanks Prof. B.-H. V. Borges (Departamento de Engenharia Elétrica, EESC, USP, Brazil) for the support with Comsol Multiphysics®. This work was financially supported by the Brazilian agencies FAPESP and CNPq under CEPOF/INOF.

8. References

Avrutsky, I. (2004). Surface Plasmons at Nanoscale Relief Gratings between a Metal and a Dielectric medium with Optical Gain. *Physical Review B*, Vol.70, No.15, pp. 155416-1 – 155416-6

Bethe, H.A. (1944). Theory of diffraction by small holes. *Physical Review*, Vol.66, No.7–8, (October), pp. 163–182.

Born, M. & Wolf, E. (1993). Principles of Optics: Electromagnetic Theory of Propagation Interference and Diffraction of Light (Pergamon), ISBN 978-008-0139-87-6, Cambridge, United Kingdom.

Bouhelier, A., Huser, T., Tamaru, T. et al. (2001). Plasmon optics of structured silver films. *Physical Review B*, Vol. 63, No. 15, (March) pp. 155404-1 – 155404-9.

Bouwkamp, C.J. (1950). On Bethe's Theory of Diffraction by Small Holes. *Philips Research Reports*, Vol.5, No.5, pp. 321–332.

Bouwkamp, C.J. (1950). On the Diffraction of Electromagnetic Waves by Small Circular Disks and Holes. *Philips Research Reports*, Vol.5, No.6, pp. 401–422.

Bouwkamp, C.J. (1954). Diffraction theory. *Reports on Progress in Physics*, Vol.17, No.1 pp. 35–100.

Chen, X. & Jiang, K. (2010). Effect of aging on optical properties of bimetallic sensor chips. *Optics Express*, Vol.18, No.2, pp.1105-1112.

Degiron, A., et al.(2004). Optical Transmission Properties of a Single Subwavelength Aperture in a Real Metal. *Optics Communications*, Vol.239, No.1-3, pp.61–66.

Ditlbacher, H., et al. (2002). Fluorescence Imaging of Surface Plasmon Fields. *Applied Physics Letters*, Vol.80, No.3, pp. 404–406.

Ditlbacher, H., et al.(2002). Two-dimensional Optics with Surface Plasmon Polaritons. *Applied Physics Letters*, Vol.81, No.10, pp. 1762–1764.

Dyankov, G., et al.(2011). Plasmon Modes Management. *Plasmonics*, Vol.6, No.4, pp. 643–650

Ebbesen, T.W., et al. (1998). Extraordinary optical transmission through sub-wavelength hole arrays. *Nature*, Vol.391, pp. 667–669.

Ferri, F.A., et al. (2011). Influence of Film Thickness on the Optical Transmission Through Subwavelength Single Slits in Metallic Thin Films. *Applied Optics*, Vol.50, No.31, pp. G11-G16.

Ferri, F.A. et al. (2012). Surface Plasmon Propagation in Novel Multilayered Metallic Thin Films. *Proceedings of SPIE*, Vol.8269, pp. 826923-1-826923-6, San Francisco, USA, January 23-26, 2012.

Fontana, E. (2006). Thickness Optimization of Metal Films for the Development of Surface-Plasmon-based Sensors for Nonabsorbing Media. *Applied Optics*, Vol.45, No.29, pp.7632–7642.

Forbes, R. & Djuric, Z (1996). Progress in understanding liquid-metal ion source operation *9th International Vacuum Microelectronics Conference*, pp.468-472, Saint Petersburg, Russia, July 7-12, 1996.

Garcia-Vidal, F.J. & Martin-Moreno, L. (2002). Transmission and Focusing of Light in One-Dimensional Periodically Nanostructured Metals. *Physical Review B*, Vol.66, No.15, pp. 155412-1 – 155412-10.

Garcia-Vidal, F.J. et al.(2010). Light Passing Through Subwavelength Apertures, *Reviews of Modern Physics*, Vol.82, No.1, pp. 729-787.

Gay, G. et al.(2006). Surface Wave Generation and Propagation on Metallic Subwavelength Structures Measured by Far-field Interferometry. *Physical Review Letters*, Vol.96, No.21, pp. 213901-1 – 213901-4.

Gay, G. et al.(2006). The Optical Response of Nanostructured Surfaces and the Composite Diffracted Evanescent Wave Model. *Nature Physics*, Vol.2, No.4, pp. 262–267.

Ghaemi, H.F., et al.(1998). Surface Plasmons Enhance Optical Transmission through Subwavelength Holes. *Physical Review B*, Vol.58, No.11, pp. 6779–6782.

Gupta, B.D. & Sharma, A.K.(2005). Sensitivity Evaluation of a Multi-layered Surface Plasmon Resonance-based Fiber Optic Sensor: a Theoretical Study. *Sensors and Actuators B:Chemical* Vol.107, No.1, pp. 40-46.

Hecht, B., et al.(1996). Local Excitation, Scattering, and Interference of Surface Plasmons. *Physical Review Letters*, Vol. 77, No. 9, pp. 1889–1892.

Huang, C.P., et al.(2007). Plasmonics:Manipulating Light at the Subwavelength Scale. *Active and Passive Electronic Components*, Vol.2007, No.30946, pp.1-13.

Hutter, E. & Fendler, J.H.(2004). Exploitation of Localized Surface Plasmon Resonance. *Advanced Materials*, Vol. 16, No. 19, pp. 1685–1706.

Jackson, J.D. (1999). Classical Electrodynamics. (John Wiley & Sons, Inc., New York, NY, 3rd edition), ISBN 978-047-1309-32-1.

Janssen, O.T.A., et al.(2006). On the Phase of Plasmons Excited by Slits in a Metal Film. *Optics Express*, Vol.14, No.24, pp.11823–11832.

Kalkum, F., et al.(2007). Surface-wave Interferometry on Single Subwavelength Slit-groove Structures Fabricated on Gold Films. *Optics Express*, Vol.15, No.5, pp. 2613–2621.

Kihm, H.W., et al.(2008). Control of Surface Plasmon Generation Efficiency by Slit-width Tuning. *Applied Physics Letters*, Vol.92, No.5, pp.051115-1 – 051115-3.

Kim, J.H. & Moyer, P.J.(2006). Thickness Effects on the Optical Transmission Characteristics of Small Hole Arrays on Thin Gold Films. *Optics Express*, Vol.14, No.15, pp.6595–6603.

Lalanne, P., et al.(2005). Theory of Surface Plasmon Generation at Nanoslit Apertures. *Physical Review Letters*, Vol.95, No 26, pp. 263902 1 – 263902-4.

Lal, S., et al.(2007). Nano-optics from Sensing to Waveguiding. *Nature Photonics*, Vol.1, No.11, pp. 641-648.

Laluet, J.-Y., et al.(2008). Generation of Surface Plasmons at Single Subwavelength Slits: from Slit to Ridge Plasmon. *New Journal of Physics*, Vol.10, No.10, pp. 105014-1 – 105401-9.

Langford R, et al.(2007). Focused Ion Beam Micro- and Nanoengineering. *MRS Bulletin*, Vol.32, No.5, pp. 417–423.

Lezec, H.J. & Thio, T.(2004). Diffracted Evanescent Wave Model for Enhanced and Suppressed Optical Transmission through Subwavelength Hole Arrays. *Optics Express*, Vol.12, No.16, pp. 3629–3651.

Li, Z.-B, et al.(2009). Fabry–Perot Resonance in Slit and Grooves to Enhance the Transmission through a Single Subwavelength Slit. *Journal of Optics A: Pure and Applied Optics*, Vol.11, No.10, pp. 1-4.

Lindquist, N.C. et al.(2012). Engineering Metallic Nanostructures for Plasmonics and Nanophotonics. *Reports on Progress in Physics*, Vol.75, No.3, pp. 1-61.

Maier, S. A. (2007). Plasmonics: Fundamentals and Applications. (Springer). ISBN 978-0387-33150-8.

Melngailis, J. (1987). Focused Ion Beam Technology and Applications. *Journal of Vacuum Science and Technology B: Microelectronics and Nanometers Structures*, Vol.5, No.2, pp.469-495.

MoberlyChan, W., et al.(2007). Fundamentals of Focused Ion Beam Nanostructural Processing: Below, at, and Above the Surface. *MRS Bulletin*, Vol.32, No.5, pp. 424–432.

Moreno, L. M. & Garcia-Vidal, F. J. (2004). Optical Transmission through Circular Hole arrays in Optically Thick Metal Films. *Optics Express*, Vol.12(16), No.16, pp. 3619-3628.

Nezhad, M., et al.(2004). Gain Assisted Propagation of Surface Plasmon Polaritons on Planar Metallic Waveguides. *Optics Express*, Vol.12, No.17, pp. 4072-4079.

Orloff, J. (1993). High-Resolution Focused Ion Beams. *Review of Scientific Instruments*, Vol.64, No.5, pp. 1-26.

Pacifici, D., et al.(2008). Quantitative Determination of Optical Transmission through Subwavelength Slit Arrays in Ag films: Role of Surface Wave Interference and Local Coupling between adjacent Slits. *Physical Review B*, Vol.77, No.11, pp. 1-5.

Palik, E.D.(1985). Handbook of Optical Constants of Solids. (Academic Press). ISBN 0125444206

Pang, Y., et al.(2007). Optical Transmission through Subwavelength Slit Apertures in Metallic Films. *Optics Communications*, Vol.280, No.1 pp. 10–15.

Porto, J.A., et al.(1999). Transmission Resonances on Metallic Gratings with very Narrow Slits. *Physical Review Letters*, Vol.83, No.14, pp. 2845–2848.

Prodan, E, et al.(2003). A Hybridization Model for the Plasmon Response of Complex Nanostructures. *Science*, Vol.302, No.5644, pp. 419–422.

Reyntjens, S. & Puers, R.(2001). A Review of Focused Ion Beam Applications in Microsystem Technology. *Journal of Micromechanics and Microengineering*, Vol.11, No.4, pp.287-300.

Rivera, V.A.G., et al.(2012). Focusing Surface Plasmons on Er 3+ Ions with Convex/Concave Plasmonic lenses. *Proceedings of SPIE*, Vol.8269, pp.82692I-1 – 8269I-6, San Francisco, USA, January 23-26, 2012.

Rivera, V.A.G., et al.(2012). Luminescence enhancement of Er 3+ Ions from Electric Multipole Nanostructure Arrays. *Proceedings of SPIE*, Vol.8269, pp.82692H-1 – 8269H-7, San Francisco, USA, January 23-26, 2012.

Sarid, D.(1981). Long-range Surface-plasma waves on very thin metal films. *Physical Review Letters*, Vol. 47, No. 26, pp. 1927–1930.

Shou, X., et al.(2005). Role of Metal Film Thickness on the Enhanced Transmission Properties of a Periodic Array of Subwavelength Apertures. *Optics Express*, Vol.13, No.24, pp.9834–9840.

Shyu, L.-H., et al.(2011). Influence of Intensity loss in the Cavity of a Folded Fabry–Perot Interferometer on Interferometric Signals. *Review of Scientific Instruments*, Vol.82, No.6, pp. 063103-1 – 063103-3.

Takakura, Y.(2001). Optical Resonance in a Narrow Slit in a Thick Metallic Screen. *Physical Review Letters*, Vol.86, No.24, pp. 5601–5603.

Tan, Y.Y., et al.(2007). Two-layered Metallic Film Induced Surface Plasmons for Enhanced Optical Propulsion of Microparticles. *Applied Physics Letters*. Vol.91, No.14, pp.141108-1-141108-3.

Tao T, et al.(1990). Focused ion beam Induced Deposition of Platinum. *Journal of Vacuum Science and Technology B: Microelectronics and nanometer structures*. Vol.8, No.6, pp.1826-1829.

Treacy, M.M.J.,(1999). Dynamical Diffraction in Metallic Optical Gratings. *Applied Physics Letters*, Vol.75, No.5, pp. 606–608.

Treacy, M.M.J.,(2002) Dynamical Diffraction Explanation of the Anomalous Transmission of Light through metallic gratings. *Physical Review B*, Vol.66, No.19. pp.195105-1–191505-11.

Volkert, C.A. & Minor, A.M.(2007). Focused Ion Beam Microscopy and Micromachining. *MRS Bulleting*, Vol.32, No.5, pp. 389-395.

Wang, J.S., et al.(1994). Tellurite glass: a new candidate for fiber devices. *Optical Materials*, Vol.3, No.3, pp.187-203

Weiner, J.(2009). The Physics of Light Transmission through Subwavelength Apertures and Aperture Arrays. *Reports on Progress in Physics*, Vol.72, No.6, pp. 1-19.

Weiner, J.(2011). The Electromagnetics of Light Transmission through Subwavelength Slits in Metallic Films. *Optics Express*, Vol.19, No.17, pp. 16139–16153.

Xie, Y., et al.(2005). Transmission of Light through a Periodic Array of Slits in a Thick Metallic Film. *Optics Express*, Vol.13, No.12, pp.4485–4491.

Ye, Y.H. & Zhang, J.Y.(2005). Enhanced Light Transmission through Cascaded Metal Films Perforated with Periodic Hole Arrays. *Optics Letters*, Vol.30, No.12, pp.1521-1523.

Yin, L., et al.(2004). Surface Plasmons at single nanoholes in Au films. *Applied Physics Letters*, Vol.85, No.3, pp. 467–469.

Zayats, A.V., et al.(2005). Nano-optics of Surface Plasmon Polaritons. *Physics Reports*, Vol.408, No.3-4, pp. 131–314.

Zynio, S.A., et al.(2002). Bimetallic layers increase sensitivity of affinity sensors based on surface plasmon resonance. *Sensors*, Vol.2, No.2, pp. 62-70.

Chapter 9

Fundamental Role of Periodicity and Geometric Shape to Resonant Terahertz Transmission

Joong Wook Lee and DaiSik Kim

Additional information is available at the end of the chapter

1. Introduction

Optical properties of the plasmonic structures based on the surface of variously structured metal plates became a subject of intense research ever since Ebbesen et al. reported the enhanced optical transmission through subwavelength hole arrays [1]. Their experimental results provided evidence that the enhanced optical properties originate from the coupling between the incident light and surface plasmon modes. Studies of such plasmonic structures and its mechanisms have a long history in the development of electromagnetism, though it has been an intense subject of research in recent years.

In the 1900s, Wood and Rayleigh reported the anomalous optical properties of reflection gratings [2,3], which are intimately connected with the excitation of surface bound waves on the metal-dielectric interfaces [4]. Hessel and Oliner presented two types of the anomalies occurring in the reflection grating surface: a Rayleigh wavelength type attributed to the usual propagating surface plasmon modes and a resonance type which is related to the guided waves supported by the grating itself [5]. On the other hand, since the 1970s, the optical properties of transmission gratings or grids have been occasionally studied in the microwave, terahertz (THz), and infrared regions [6-10]. In particular, Lochbihler and Depine in 1993 presented a theoretical approach for calculating the fields diffracted by the gratings made of highly conducting wires [11]. Here they calculated Maxwell's equations based on modal expansions inside the metal and Rayleigh expansions outside the metal, applying the theoretical treatments by means of surface impedance boundary condition method [11,12].

This research field reinvigorated after the Ebbesen's pioneering work which was focused on the understanding of mechanisms of the enhanced transmission through periodically perforated holes or slits on metallic plates. The one-dimensional plasmonic structures with periodic arrays of slits have been studied by some groups with theoretical and experimental approaches in the optical region [13-18]. These results strongly support the existence of two mechanisms of generating enhanced optical transmission: the geometric shape controlled mainly by the slit thickness and the periodicity determining the excitation of coupled surface plasmon polaritons (SPPs) on both surfaces of the perforated metallic structures. In particular the former called as a Fabry-Perot-like behavior was studied theoretically in a single narrow slit and clearly verified by strong resonant transmission of microwave radiation [19,20].

On the other hand in two-dimensional plasmonic structures with periodic arrays of holes, the geometric shape effect has been underestimated in spite of many researches about the strong transmission enhancement in the optical region [21-26]. Recently, the geometric shape resonance however was emerged as one of the dominant mechanisms of the strong transmission enhancement [27-29]. In particular, Koerkamp et al. already noted that the strong optical transmission is strongly influenced by a hole shape and a shape change from circular to rectangular [30]. The importance of the geometric shape was also confirmed by studies of strong enhancement of the light transmission through random arrays of holes [30-32] and single apertures [33-36], in particular a single rectangular hole theoretically studied by Garcia-Vidal et al. [37].

Unlike most typical systems in optical and infrared regions, THz time-domain spectroscopy system based on coherent THz radiation is ideally suited for phase-sensitive, broadband transmission measurements [38-41]. Furthermore, long wavelengths in THz region, 3000 μm to 30 μm (0.1 THz to 10 THz) in general, permit relatively simple fabrication of free-standing samples which can exclude complex surface modes by the substrate. For these reasons, the coherent THz waves have been generally used to study SPP behaviors [42-55]. The SPPs in the strictest sense however do not exist on perfectly conducting metal-dielectric interfaces since surface electromagnetic fields on the dielectric side cannot be strongly confined to the interface. However, recently the concept of the SPP has been extended even to include perfectly good metals with corrugations, which is referred to simply as designer (or spoof) SPP [56-58]. Pendry et al. reported the electromagnetic surface excitations localized near the surface which are governed by an effective permittivity of the same plasma form [56].

In the THz region, the most important thing is that the perfect transmission, of up to near-unity, at specific frequencies can be achieved since most metals become perfect conductor and therefore have extremely small ohmic loss fraction. Until now, a few groups have theoretically predicted the perfect transmission [59-64] and experimentally observed the near-unity transmission [65-67] in the plasmonic structures with periodic arrays of slits and holes. However, despite great progresses in realizing the perfect transmission, we do not have the clear understanding of what are the relative contributions of the geometric shape and the periodicity effects toward the perfect transmission in specific conditions.

Here, we first report on the importance of the combined effects of the geometric shape and periodicity in enhancing THz transmission through one-dimensional periodic arrays of slits. Theoretical predictions based on perfect conductor model show that the measured transmission peaks lie within the broad geometric shape band. We also report that the perfect transmission can be realized by the geometric shape resonance appearing in spectral region below first Rayleigh minimum. Calculated near-field distributions confirm that the perfect transmission is caused by the concentration of the electric and magnetic fields within the silts. Furthermore, we find that, despite the existence of only the geometric shape effect in random arrays of slits of relatively thick thicknesses, the enhanced transmission comes to the perfect transmission in the long wavelength region. Angle dependent transmission spectra and relative phase measurements clarify the relative roles of the geometric shape and periodicity toward the perfect transmission.

2. Experimental setup

In our experiments, we used a standard THz time-domain spectroscopy system with a spectral range from 0.1 to 2.5 THz, based on a femtosecond Ti:sapphire laser which generates optical pulses with average power of 600 mW and temporal width of about 120 fs set at a center wavelength of 760 nm. Experimental setup is shown in Fig. 1. A photoconducting antenna method is used to generate THz pulses. A crystal for the emitter is undoped, <110> oriented semi-insulating GaAs (INGCRYS Laser System Ltd.) with dark resistivity in range 5.3×10^7 to 5.6×10^7 Ohm cm. Two metallic electrodes are painted by silver paint on one side of the crystal, having a separation of about 0.4 mm which is among a large aperture emitter. Unlike a typical biased system of the photoconductive antenna, an ac bias voltage of 300 V and 50 kHz square wave was applied to the electrodes of the crystal [68]. To make this system, we used a high voltage pulse generator (DEI Model PVX-4150) which generates the squared high-voltage sources by combining DC high voltage source from a power supply (F. u. G. Model MCL 140-650) with square function generated using a pulse/function generator.

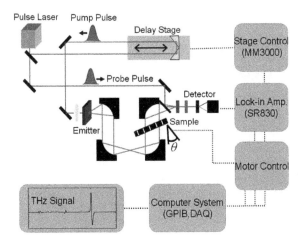

Figure 1. Schematic diagram of the THz time-domain spectroscopy setup. The angle dependent transmission spectra are measured by tilting the sample stage.

Optical pulses generated from the femtosecond Ti:sapphire laser are divided into two parts, pump and probe beams, by a beamsplitter. The pump beam is focused on the surface of the emitter by using a 25 mm focal lens and immediately generates the THz wave which is essentially single cycle electromagnetic pulses. The generated THz wave is collimated by using an off-axis parabolic mirror at a focal distance. The degree of collimation is of the order of $\Delta\theta \sim \lambda/D \sim 1/100$ radian because our beam size D is about 5 cm and the THz wave is a point-like source. The collimated THz wave with the beam diameter of about 5 cm impinges on a reference box or samples. The angle-dependent transmission spectra are measured by tilting the sample stage in range -10 to 50 degrees. The THz pulses transmitted through the samples meet the optical probe beam again on an electro-optic crystal (ZnTe). The polarization change of the optical probe beam due to the Pockels effect is detected by a home-built differential photo-detector and provides us information of both amplitude and phase of the THz wave [69,70]. The signals are automatically acquired by a synchronized system (as shown in Fig. 1) of three components, a motion controller which tilts the sample stage, a data acquisition system, and a delay stage which controls time delay between pump and probe optical pulses. The whole setup is also enclosed in a box which is purged with dry nitrogen gas to reduce the absorption effects of water vapor.

We performed our experiments as follows. First, the collimated THz wave impinges on a reference box made of a 2 cm by 2 cm square hole punctured on a piece of aluminum. We measure the transmission spectrum through the reference box. We then mount our samples with the plasmonic structures of periodic or random arrays of slits perforated on aluminum plates, right on top of the reference box, and measure the transmission spectrum. Our system provides the time-resolved trace of the transmitted THz electric field with subpicosecond temporal resolution as an original signal. The fast Fourier transforms of the measured time-domain THz pulses give us information on both spectral amplitude and phase. The normalization is carried out over the entire frequency simply by dividing the spectral amplitude of the transmitted THz pulse with one of the reference beam transmitting through the reference box.

3. Sample fabrication

The samples were fabricated by a micro-drilling and a femtosecond laser machining methods. The latter is based on laser ablation which is performed by amplified femtosecond pulses. The pulses have energy of up to 1 mJ centered at 800 nm and a repetition rate of 1 kHz. The positions of the focused laser pulses are accurately controlled by using a galvanometer scanner (Scanlab AG, Germany) and a mechanical shutter system. The line edge roughness of the perforated slits is controlled to make successful slit structures, having the LER 3σ value of less than 3 µm which is two orders of magnitude less than the THz wavelengths of interest. The achieved high quality originates from the superiority of femtosecond laser machining system. The samples manufactured by this method have a fixed period d of 500 µm and different thicknesses h of 17, 50, 75 and 153 µm respectively. The silt widths a of these samples are 78, 80, 85, and 83 µm respectively, which are properly designed to be subwavelength [as shown in Fig. 2(a)].

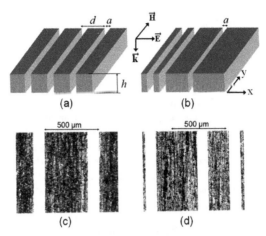

Figure 2. (a), (b) Schematics of the periodic and random arrays of slits of width a, period d, and thickness h, respectively. (c), (d) Microscopic images of the periodic and random arrays of 400 μm thick slits, respectively (Fig. (1) in reference [71]).

On the other hand, the samples with thicknesses of 200 and 400 μm were fabricated by the micro-drilling method by using a screw with a diameter of 100 μm which is a minimum size in diameter of available and commercial screw. These thick plates were chosen for explicitly observing Fabry-Perot-like modes inside the slits. In particular the samples with thickness of 400 μm were considered suitable for randomization of the slit structures, based on the different frozen-phonon fashion of neighboring silts [as shown in Fig. 2(b)]. The position of the ith slit by the frozen-phonon fashion is given by

$$x_i = x_0 + id + \Delta x_i \qquad (1)$$

where x_0 is the position of the basis of coordinate and Δx_i is the random amount of the ith slit which is getting by generating random values [72]. We displaced the position of each slit with random amounts of ± $d/4$ and ± $d/2$ which, in this chapter, are called as *Random Type I* and *Random Type II*.

Comparison between transmission properties through random and periodic arrays of slits reveal that the transmission spectra strongly affected by the geometric shape show similar spectral waveforms in both random and periodic arrays, since the geometric shape resonance depends on intrinsic properties of the structure factors not the structural arrangement of the slits. The geometric shape in the one-dimensional plasmonic structures is determined by the thickness h of the slits, providing the Fabry-Perot-like modes $f_c = nc/2h$ where c is the speed of the light in vacuum and n is the integer mode index. In contrast, the periodicity effect might be found in transmission spectra through the periodic arrays not random, having different constructive interference patterns and different far-field transmissions. Therefore the relative contribution of the geometric shape and periodicity is directly compared in this chapter.

4. Theoretical framework

Herein, we are first interested in theoretical approach of THz electromagnetic wave transmission through the periodic arrays of slits, which may allow us to improve our intuitive understanding of the underlying physics and to find the mechanisms enhancing the transmission. To calculate the transmittance, the metallic structures are considered as a perfect conductor so that finite conductivity effects are neglected. This is a good approximation in the THz region, since the real and imaginary parts of dielectric constant of most metals are of the order of -30000 and 100000 respectively. We also consider a simple model system with rectangular shaped slits as shown in Fig. 2(a). In this system, the slits are along the y axis and a direction of propagation of the incident THz wave is the z axis which is normal to the sample surface. The polarization of the incident plane wave is placed along the x axis (TM polarization). Here, there are three different regions: the incident and reflection region (region I), the metal and perforated slit region (region II), and the transmission region (region III).

Let $f(x,z)$ be the spatial function of the component along the z axis of the magnetic field. The fields can be expressed in terms of Rayleigh expansions in region I and III and modal expansion in region II [11,46]. The boundary matching is only carried out at the metal-dielectric interfaces of I-II and II-III since which are semi-infinite. Here we use the single mode approximation inside the slits since the slit width is much smaller than the wavelength. The magnetic fields are described as

$$f_1(x,z) = \sum_{-\infty}^{\infty} (R_n e^{i\chi_n(z+h/2)} + \delta_{n0}e^{-ik(z+h/2)})e^{i\alpha_x x}, \text{ region I} \tag{2}$$

$$f_3(x,z) = \sum_{-\infty}^{\infty} T_n e^{-i\chi_n(z-h/2)}e^{i\alpha_x x}, \text{ region III} \tag{3}$$

and

$$f_2(z) = A\frac{\sin(kz)}{\sin(kh/2)} + B\frac{\cos(kz)}{\cos(kh/2)}, \text{ region II} \tag{4}$$

where $\alpha_n = 2\pi n/d$, $\chi_n = \sqrt{k^2 - \alpha_n^2}$, $k = \omega/c = 2\pi/\lambda$, and R_n and T_n are the complex amplitudes of the reflected and transmitted diffracted waves respectively. The continuity of the tangential components of the fields implies the continuity of $f(x,z)$ and its normal derivative at $z = \pm h/2$ along $0 \leq x \leq a$. These conditions are

$$A + B = \sum_{n=-\infty}^{\infty} (R_n + \delta_{n0})e^{i\alpha_n x} \tag{5}$$

for the continuity of the field at $z = -h/2$,

$$Ak\cot(\frac{kh}{2}) - Bk\tan(\frac{kh}{2}) = \sum_{n=-\infty}^{\infty} (i\chi_n R_n - ik\delta_{n0})e^{i\alpha_n x} \tag{6}$$

for the continuity of the normal derivative of the field at $z = -h/2$,

$$-A + B = \sum_{n=-\infty}^{\infty} T_n e^{i\alpha_n x} \tag{7}$$

for the continuity of the field at $z = h/2$, and

$$Ak\cot(\frac{kh}{2}) + Bk\tan(\frac{kh}{2}) = \sum_{n=-\infty}^{\infty} (-i\chi_n T_n)e^{i\alpha_n x} \tag{8}$$

for the continuity of the normal derivative of the field at $z = h/2$.

With the orthogonality, we apply some of integral equation to obtain simple solution with proper projection and integration. Multiplying a term $e^{-i\alpha_m x}$ and integrating its results of Eqs. (5)-(8) give

$$A + B = \sum_{n=-\infty}^{\infty} (R_n + \delta_{n0})P_n \tag{9}$$

$$\left[Ak\cot(\frac{kh}{2}) - Bk\tan(\frac{kh}{2})\right]Q_n = i\chi_n R_n - ik\delta_{n0} \tag{10}$$

$$-A + B = \sum_{n=-\infty}^{\infty} T_n P_n \tag{11}$$

and

$$\left[Ak\cot(\frac{kh}{2}) + Bk\tan(\frac{kh}{2})\right]Q_n = -i\chi_n T_n \tag{12}$$

where the quantities P_n and Q_n appearing in Eqs. (9)-(12) are defined as

$$P_n = \frac{1}{a}\int_0^a e^{i\alpha_n x}dx \tag{13}$$

and

$$Q_n = \frac{1}{d}\int_0^a e^{-i\alpha_n x}dx \tag{14}$$

respectively. By using Eqs. (9)-(12), we obtain

$$A = \frac{P_0}{1 + ikW\cot(kh/2)} \tag{15}$$

$$B = \frac{P_0}{1 - ikW \tan(kh/2)} \tag{16}$$

$$R_n = 1 - \frac{2ikQ_nP_0}{\chi_n} \left[\frac{\cot(kh) - ikW}{1 + k^2W^2 + 2ikW\cot(kh)} \right] \tag{17}$$

and

$$T_n = \frac{2ikQ_nP_0}{\chi_n \sin(kh)} \left[\frac{1}{1 + k^2W^2 + 2ikW\cot(kh)} \right] \tag{18}$$

where W is defined as

$$W = \sum_{n=-\infty}^{\infty} \frac{P_nQ_n}{\chi_n} \tag{19}$$

For normal incidence, zero-order transmittance for the periodic arrays of slits is expressed as follows

$$T_0 = \frac{2ia}{d\sin(kh)} \left[\frac{1}{1 + k^2W^2 + 2ikW\cot(kh)} \right] \tag{20}$$

5. Experimental results

5.1. Enhanced transmission in thin metal plates

Theoretical calculations are first compared with experimental results of samples with the thicknesses of 17, 75 and 153 μm, which are carefully chosen so that the peak positions due to the geometric shape resonance are placed on the higher spectral region than the first Rayleigh minimum. Measured THz time traces and corresponding spectral amplitudes obtained by the fast Fourier transform method are shown in Fig. 3(a) and 3(b) respectively. The top curves in Fig. 3(a) and 3(b) show the reference signals transmitted through the reference box. The single cycle THz pulses with a temporal width of about 2 ps in time domain are transformed to the spectral pulses with a well-formed shape centered at 0.5 THz and a broad spectral width. In our experimental setup, we effectively removed multiple reflections of the THz pulses occurring from the emitter, the electro-optic crystal and a beamsplitter, which can help assure good quality of the measured time-resolved THz signals.

The thinnest sample with the thickness of 17 μm does not have the peak position of the enhanced transmission locating inside the presented spectral region. However, the two curves for the thicknesses of 75 and 153 μm in Fig. 3(a) have significant long-time oscillations with the periods of 0.87 and 1.80 picoseconds respectively, which also appears in the Fourier transformed spectra as peaks at 1.14 and 0.56 THz respectively (Fig. 3(b), gray arrows in bottom two curves). Here the peak positions of the enhanced transmission appear at the frequency regions slightly below the Rayleigh minima $f_R = c/d$, which indicate that the transmission properties are related to the periodicity effect. On the other hand, the peak positions shift toward longer wavelength region as the sample thickness increases, keeping the typical characteristics of Fabry-Perot-like resonance. These observations indicate that the enhanced transmissions are attributed to the combined effects of the geometric shape resonance and the periodicity.

Theoretical calculations shown in Fig. 3(c) are in well agreement with the experimental results except for extremely sharp transmission peaks immediately below the first Rayleigh wavelength. In particular in the cases of the two thinnest samples, the peak linewidths have almost infinitesimal values less than several gigahertz, which cannot be experimentally observed because of the finite temporal range of time domain signals and finite sample size. Theoretically predicted peak suddenly appears at 0.56 THz for h=153 μm as shown in bottom curve of Fig. 3(b). This appearance results from the approach of the geometric shape resonance toward the first Rayleigh minimum at which the first diffracted orders become evanescent.

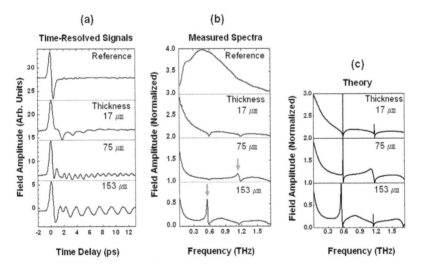

Figure 3. (a) Time traces of the incident terahertz wave (the upper figure) and the transmitted waves for three samples with a fixed period of 500 μm and different thicknesses of 17, 75, and 153 μm. The slit widths of the three samples are 78, 80, and 83 μm respectively. (b) Fourier transforms for four time traces in (a). The gray arrows represent the resonant peak positions. (c) Theoretical calculations for three samples in (b).

The independent contribution of Fabry-Perot-like resonance can draw out from single-slit approximation which can be achieved by carrying the distance between slits to be infinite. Figure 4(a) and 4(c) show the resonant peaks of transmission spectra through the samples with the thicknesses of 75 and 153 μm respectively, theoretically predicted by the single-slit approximation. The red-shifted Fabry-Perot conditions predicted by Takakura in a single narrow slit of a thick metallic plate assure the precision of the theoretical predictions. These peak positions induced only by the Fabry-Perot-like resonance are compared with the measured angle-dependent transmission amplitudes [Fig. 4(b) and 4(d)], since the geometric shape resonance is angle-independent. For the case of the 75 μm thickness, the experimental results show a resonance band over all incident angles around 1.3 THz, which is reasonably predicted by the first Fabry-Perot-like resonance band centered at 1.2 THz [Fig. 4(a), shadowed area]. The theoretical prediction and experimental results for the sample of the 153 μm thickness also show the corresponding two spectral bands centered at 0.65 and 1.45 THz. From these theoretical prediction and experimental results, we note that the peak positions of resonant transmission can be experimentally determined by broad Fabry-Perot-like resonance band existing nearby.

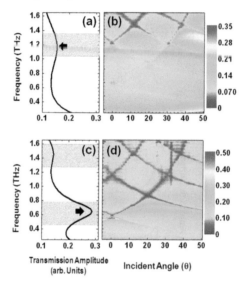

Figure 4. (a), (c) Transmission spectra, at normal incidence, predicted by single-slit approximation for the samples with the thicknesses of 75 and 153 μm respectively. Arrows represent the red-shifted first Fabry-Perot positions calculated by Takakura (Eq. (9), ref. [19]). (b), (d) Measured angle-dependent transmission spectra for the samples with the thicknesses of 75 and 153 μm respectively.

At this point, we can expect that perfect transmission phenomenon will be appeared under the condition of overlapping between the Fabry-Perot-like resonance band and totally evanescent spectral regions below the first Rayleigh wavelength. The first diffracted waves touch down on the metal surface at the wavelength of the first Rayleigh minimum, at which wavelength eventually all higher diffracted orders become evanescent which is the reason why the total incident energy can be converged to the zero-order transmission amplitude. We therefore design a thicker sample to realize the perfect transmission.

5.2. Perfect transmission in thick metallic plates

The sample with the thickness of 200 μm eventually has broad Fabry-Perot-like resonance band centered at 0.45 THz which is located in the spectral region below the first Rayleigh minimum appearing at 0.6 THz as shown in Fig. 5(a). Angle dependent transmission amplitudes [Fig. 5(b)] show the perfect transmission (in this sample, over 99% at normal incidence) centered at 0.5 THz. Despite the two transmission minima lines caused by Rayleigh frequencies for the ±1 diffraction orders, the perfect transmission appears at a broad incident angle range up to 10° because of the strong contribution of the Fabry-Perot-like resonance. Theoretical calculation with Eq. (20) is in good agreement with experimental result at normal incidence.

The perfect transmission phenomenon necessarily appears when the peak position of the enhanced transmission is located anywhere in the spectral region below the first Rayleigh frequencies, since all the diffracted orders of the transmittance T_n except for zeroth order become zero. A very useful theoretical condition for the perfect transmission can be therefore derived by setting the zeroth order reflection coefficient at zero as following

$$|R_0| = \left| 1 - \frac{2iQ_0P_0}{1}\left[\frac{\cot(kh) - ikW}{1 + k^2W^2 + 2ikW\cot(kh)} \right] \right| = 0 \tag{21}$$

Below the first Rayleigh frequencies, W can be considered as

$$W = \frac{P_0Q_0}{k} + \frac{i\gamma}{k} \tag{22}$$

since all the reflected diffracted orders except for zeroth order become evanescent. Where γ is an imaginary part of kW. We can then write the condition of the perfect transmission as

$$1 - (P_0Q_0)^2 - \gamma^2 - 2\gamma\cot(kh) = 0 \tag{23}$$

For the theoretical calculation of the first Fabry-Perot-like resonance centered at 0.48 THz [Fig. 5(c)], a magnetic field profile inside the slits becomes symmetric [Fig. 5(d)] which is proved by the condition for the perfect transmission given by Eq. (23). Indeed, the magnetic field inside the slits given by Eq. (4) converges into the symmetric waveform at the first Fabry-Perot-like resonance as

$$f_2(z) \cong B\frac{\cos(kz)}{\cos(kh/2)} \tag{24}$$

since a denominator term of B is minimized under the condition of Eq. (23). It can be also intuitively understood that, under the condition of the perfect transmission at the first Fabry-Perot-like resonance, the electric field strengths at both sides of entrance and exit of the slits are the same despite opposite orientations of the electric fields at each position. Because the light coupling between resonant modes of the structured metal surfaces in the transmission and reflection regions is related to the matching condition of the phase of the electric and the magnetic fields, the electric field strength at the entrance of the slits cannot be reproduced by the asymmetric profiles shown in Fig. 5(d).

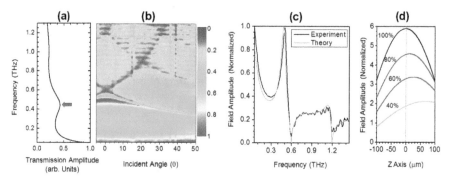

Figure 5. (a) Transmission spectra, at normal incidence, predicted by single-slit approximation for the sample with the thickness of 200 μm. An arrow represents the peak position of first Fabry-Perot-like resonance mode. (b) Measured angle-dependent transmission spectra for the sample with the thickness of 200 μm. The period is 500 μm and the slit width is 100 μm. (c) Experimental and theoretical results, at normal incidence, of the sample with the thickness of 200 μm. (d) The magnetic field profiles along the z direction inside the slits calculated by the perfect conductor model. Under the condition of the perfect transmission appearing at the first Fabry-Perot-like mode, the field profile is symmetric (black line). At neighboring frequencies with the transmission amplitudes of 80 (blue line), 60 (red line) and 40% (green line) respectively, the field profiles become asymmetric.

Near-field distributions of THz waves above the metal surfaces can also give a key for understanding the perfect transmission phenomenon. The enhancement factor of the near electric fields onto the slits can be intuitively predicted by using the language of the surface impedance as

$$Z = \frac{a}{d} \frac{E_{slit}}{H_0} + \frac{(d-a)}{d} \frac{E_{metal}}{H_0} \tag{25}$$

where E_{slit}, E_{metal} and H_0 are the electric fields along the x direction at the slits and the metal and the incident magnetic field respectively. Since the effective surface impedance becomes one under the condition of the perfect transmission and there is no tangential component of the electric field on the metal surface, we can write Eq. (25) as

$$Z_{eff} = \frac{a}{d} \frac{E_{slit}}{H_0} = 1 \tag{26}$$

The degree of the electric field enhancement is not only proportional to the incident magnetic field, but also inversely proportional to the areal sample coverage $\beta = a/d$ as

$$E_{slit} = \beta^{-1} H_0 \tag{27}$$

For the case of the areal sample coverage of $\beta = 0.2$ ($a = 100$ and $d = 500$ μm), the theoretical results show the enhancement factor of 5 on the entrance of the silts as shown in Fig. 6(a). At a specific frequency at which the perfect transmission is realized, the incident waves are focused on the entrance of the slits due to the strong evanescent waves built up by a touchdown of all orders of diffracted modes on metal surface, followed by coupled with the Fabry-Perot-like resonant modes inside the slits, and then the focused waves reached at the exit of the slits are completely converted to the transmission.

These results suggest a possibility of realizing even stronger enhancement of the electric field strength onto the slits. We therefore fabricated a sample with the areal sample coverage of $\beta = 0.1$ by increasing the period. (Reducing the slit width is actually more effective, but the technical limit of the micro-drilling method to fabricate thick metallic plates is 100 μm.) Calculated near-field distribution of the sample is given by Fig. 6(b), showing the enhancement factor of 10 corresponding to the inverse of the areal sample coverage. The near-field enhancement necessarily induces the transmission enhancement into the far-field. Measured angle dependent transmission spectra shown in Fig. 7 show not only angle-independent, broad spectral bands near 0.3 and 0.6 THz caused by the geometric shape resonance, but also a strong transmission peak with the value of about 0.85 at 0.26 THz at normal incidence. Though the peak value does not reach the THz transparency because of a narrow spectral bandwidth, the enhancement factor compared with the areal sample coverage of 0.1 is over eight. Controlling the areal sample coverage by using the period or width is therefore extremely important in realizing a high power near-field THz source.

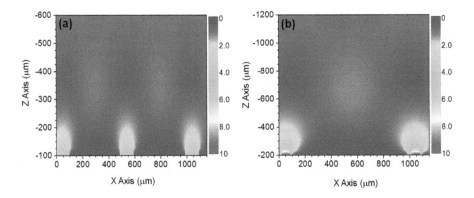

Figure 6. Electric field amplitudes for the samples with different periods of 500 (a) and 1000 μm (b) respectively. The thickness h is 400 μm and the width a is 100 μm in both cases. Enhancement factors at the entrance of the slits are 5 and 10 respectively.

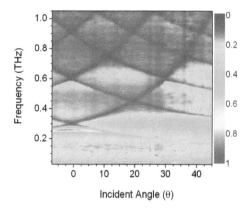

Figure 7. Measured angle-dependent transmission spectra for the sample with the period of 1000 μm. The thickness is 400 μm and the slit width is 100 μm.

5.3. THz transparency by geometric shape

At this point, it is important to understand that what the relative contributions of the periodicity and the geometric shape resonance toward the perfect transmission are in a given situation. From the experimental viewpoint, the best method to compare the difference between the relative contributions of two mechanisms is to fabricate the samples with random arrays and to investigate its transmission properties. The calculated transmission spectra as shown in Fig. 8 play an important role in determining the appropriate sample parameters. For the convenience of comparison, the transmission spectra are plotted versus the first Fabry-Perot frequency, $f_c = c/2h$ instead of the slit thickness.

In the cases of the relatively thinner metal plates, there are two peak maxima of the transmission spectra: one corresponds to the first Fabry-Perot-like mode, $f_c = c/2h$, and the other corresponds to the so-called zeroth Fabry-Perot-like mode appearing at the long wavelength region near 0 THz, which is originated from the geometric shape effect of the semi-infinite slits in a thin metal film. In the long wavelength region near 0 THz, an aluminum plate with the thickness of several hundreds of microns can be actually regarded as an optically thin metal film. With increasing the thickness of the aluminum plates, the peak maxima originating from higher Fabry-Perot-like modes such as $f_c = 2c/2h$, $3c/2h$, and etc are added consecutively, maintaining the perfect transmission.

All curves of the Fabry-Perot-like modes show that the curves approach to straight lines of the classical Fabry-Perot positions (Fig. 8, dotted lines) as getting away from the wavelength of the first Rayleigh minimum of 0.6 THz. The calculated results may therefore present that, when the peak maxima approach to the long wavelength limit, the contribution by the coupled evanescent surface modes presented in Eq. (19) becomes weak. In our experiments, we therefore fabricated the samples of periodic and random arrays of slits with the thickness of 400 μm since in which there are three different peak maxima originating from the zeroth

(Fig. 8, C), first (Fig. 8, B) and second (Fig. 8, A) Fabry-Perot-like modes and each peak maximum may be stimulated by different contributions of the periodicity and the geometric shape. In fact, as the peak position is getting away from the wavelength of the first Rayleigh minimum toward longer wavelength region, the geometric shape effect becomes dominant.

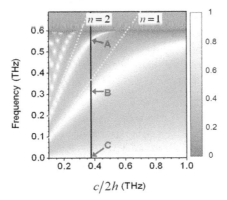

Figure 8. Calculated transmission spectra plotted versus the frequency of the first Fabry-Perot-like resonance. The dotted lines represent the first and second classical Fabry-Perot resonances. The sample with the thickness of 400 μm has three Fabry-Perot resonance modes (A, B, and C) appearing at the frequency region below the first Rayleigh minimum.

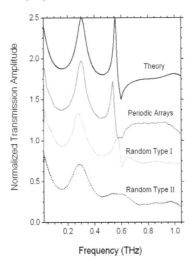

Figure 9. Normalized transmission spectra at normal incidence for three types of the samples: periodic arrays (dark gray line), random type I (bright gray line), and random type II (dotted line). The black line is the calculated transmission spectra for the case of the periodic arrays. These samples have same areal sample coverage of 0.2 and same thickness of 400 μm. The randomization of the slit structures is based on the different frozen-phonon fashion.

Figure 9 shows the normalized transmission amplitudes of periodic and random arrays of silts at normal incidence. Here the samples have same slit width a of about 100 μm and same areal sample coverage of β=0.2. As predicted in theoretical calculations, the calculated transmission spectrum (Fig. 9, a solid line) shows three peak maxima at 0.56, 0.30 and near zero THz, corresponding to the second, first and zeroth Fabry-Perot-like modes respectively. The measured transmission spectrum (Fig. 9, a dark gray line) of the periodic arrays of slits agrees well with the calculated one, except slight disagreement at the peak maximum of the second Fabry-Perot-like mode. Here the experimental linewidth of the second Fabry-Perot-like mode does not completely reproduce the theoretical one since which is too sharp to experimentally observe in real sample.

We then consider two random structures with the different frozen-phonon fashion of neighboring slits. The Random Type I and II samples were fabricated by displacing the position of each slit of periodic arrays with random amounts of $\pm d/4$ and $\pm d/2$ respectively. The transmission spectrum of the Random Type I (Fig. 9, a bright gray line) shows a slightly lower peak value at 0.30 THz and a considerably decreased peak value at 0.56 THz, but overall the spectral shape of the transmission spectra are well matched, particularly reproducing the peak value due to the zeroth Fabry-Perot-like mode. The measured transmission spectrum of the Random Type II (Fig. 9, a dotted line) clearly exhibits the spectral properties of the perfectly random arrays of slits, which does not show any more the transmission minimum at 0.6 THz originating from the first Rayleigh minimum due to the periodicity effect only. On the other hand, although the peak value appearing near 0.3 THz is slightly decreased with further increasing the fraction of random amounts, the spectral shape and peak position can be not only considered unchangeable, the resonant peak also exhibits an enhancement factor larger than 3.5. Furthermore, the resonant peaks of the zeroth Fabry-Perot-like mode are more dramatic where all the transmission spectra in theory and experiments are perfectly matched, showing the transmittance more than 80%.

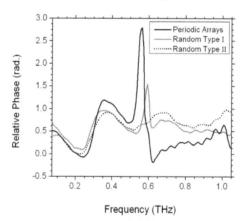

Figure 10. Relative phase changes in radians at normal incidence for three types of the samples: periodic arrays (black line), random type I (gray line), and random type II (dotted line).

The relative phase of transmitted THz waveforms not only improves the understanding for the relative contributions of the geometric shape and the periodicity but also makes more clearly what the dominant mechanism in enhancing the THz transmission is. When the geometric shape effect becomes dominant, there will be a little difference between the random and periodic arrays of slits, as shown in Fig. 10. In addition, the geometric shape does not depend on the incident angle of the THz waves. Shown in Fig. 11 are the angle dependent transmission amplitudes for the samples. For the sample of periodic arrays of slits, there are two dominant transmission minima lines crossing at 0.6 THz at zero angle which is the Rayleigh wavelengths corresponding to the ±1 diffracted orders as

$$f_{\pm 1}(\theta) = \frac{c}{d[1 \mp \sin(\theta)]} \qquad (28)$$

The Rayleigh wavelengths determined by the grating period distinctly appear as transmission minima lines over all the incident angle of the THz waves, generating an angle-dependent and strongly enhanced transmission peak near 5.4 THz. With increasing the randomness, the transmission minima lines of the Rayleigh wavelengths become indistinct as shown in Fig. 11(b) and eventually disappear as shown in Fig. 11(c). Similarly, the angle dependent transmission peak appearing at the sample of periodic arrays of slits also disappears since its strong enhancement is caused by the strong field accumulation of the evanescent surface waves by the periodic structures. On the contrary, the enhanced peaks appearing near 0.3 THz are essentially angle independent, making the strong enhancement.

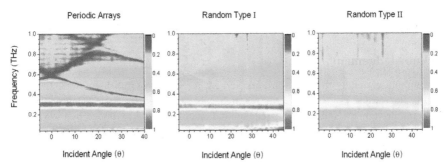

Figure 11. Normalized angle dependent transmission amplitudes for three types of the samples: periodic arrays (a), random type I (b), and random type II (c).

6. Conclusion

In conclusion, we have demonstrated two mechanisms for perfect transmission appearing at specific frequencies through the periodic and random arrays of slits. The theoretical results show that the perfect transmission can be realized in spectral region below the first Rayleigh minimum determined by the periodicity. Under the condition for the perfect transmission, the symmetric electric- and magnetic-field profiles inside the slits are excited and the incident energy is funneled onto the slits with the enhancement factor proportional to the inverse of the areal sample coverage. The measured transmission spectra through the slits of thick metal plates show the enhanced transmission of perfect transmission, which is attributed by the increase of the spectral linewidth resulting from the approach of the geometric shape resonance to the spectral region below the first Rayleigh minimum. The resonant bands of the geometric shape presented by the single silt approximation and the measured angle dependent transmission spectra not only are angle independent but also determine the spectral peak positions of the perfect transmission.

The random arrays of slits designed to remove the effect of the periodicity clearly show the dominant contribution of the geometric shape resonance toward the enhanced transmission. In the long-wavelength region far from the first Rayleigh minimum, the geometric shape resonance becomes dominant and the transmission enhancement factors are almost equal to the cases of the periodic arrays of slits. The strong transmission and local field enhancements open a possibility of potential applications in perfect transmission of solid materials such as semiconductors and biological materials requiring the high power THz sources and in designing photonic devices such as transmission filters, low-pass frequency filters, and transmissive waveguides requiring high transmission at desired frequencies.

Author details

Joong Wook Lee*
Advanced Photonics Research Institute, GIST, Gwangju, Republic of Korea

DaiSik Kim
Department of Physics and Astronomy, Seoul National University, Seoul, Republic of Korea

Acknowledgement

The authors gratefully thank Q. H. Park and M. A. Seo for useful discussions and S. C. Jeoung for sample preparation. This research was supported by Basic Science Research Program through the National Research Foundation of Korea (NRF) funded by the Ministry of Education, Science and Technology (2010-0021181) and by the APRI Research Program of GIST.

7. References

[1] Ebbesen T. W., Lezec H. L., Ghaemi H. F., Thio T., and Wolff P. A. (1998) Extraordinary optical transmission through sub-wavelength hole arrays. Nature 391: 391, 667-669.
[2] Wood R. W. (1902) On a remarkable case of uneven distribution of light in a diffraction grating spectrum. Phil. Mag. 4: 396-408.
[3] Rayleigh L. (1907) On the passage of electric waves through tubes, or the vibrations of dielectric cylinders. Phil. Mag. 14: 60-65.

* Corresponding Author

[4] Fano U. (1941) The theory of anomalous diffraction gratings and of quasi-stationary waves on metallic surfaces (Sommerfeld's waves). J. Opt. Soc. Am. 31: 213-222.

[5] Hessel A. and Oliner A. A. (1965) A New Theory of Wood's Anomalies on Optical Gratings. Appl. Opt. 4: 1275-1297.

[6] Ulrich R. (1968) Interference filters for the far infrared. Appl. Opt. 7: 1987-1996.

[7] Chen C. C. (1970) Transmission Through a Conducting Screen Perforated Periodically with Apertures. IEEE trans. Microwave Theory Tech. 18: 627-632.

[8] Bliek P. J., Botten L. C., Deleuil R., Paedran R. C. M., and Maystre D. (1980) Inductive grids in the region of diffraction anomalies: theory, experiment, and applications. IEEE trans. Microwave Theory Tech. 28: 1119-1125.

[9] Keilmann F. (1981) Infrared high-pass filter with high contrast. Int. J. Infrared Millimeter Waves 2: 259-272.

[10] Ulrich R. (1974) Modes of propagation on an open periodic waveguide for the far-infrared, in optical and acoustical micro-electronics, ed. by J. Fox, Micr. Res. Inst. Symp. Series 23: Polytec. Press, New York 359-376.

[11] Lochbihler H. and Depine R. (1993) Highly conducting wire gratings in the resonance region. Appl. Opt. 32: 3459-3465.

[12] Lochbihler H. (1994). Surface polaritons on gold-wire gratings. Phys. Rev. B 50: 4795-4801.

[13] Porto J. A., Garcia-Vidal F. J., and Pendry J. B. (1999) Transmission resonances on metallic gratings with very narrow slits. Phys. Rev. Lett. 83: 2845-2848.

[14] Cao Q. and Lalanne P. (2002) Negative role of surface plasmons in the transmission of metallic gratings with very narrow slits. Phys. Rev. Lett. 88: 057403.

[15] Collin S., Pardo F., Teissier R., and Pelouard J. L. (2002) Horizontal and vertical surface resonances in transmission metallic gratings. J. Opt. A: Pure Appl. Opt. 4: 154-160.

[16] Astilean S., Lalanne P., and Palamaru M. (2000) Light transmission through metallic channels much smaller than the wavelength. Opt. Commun.175: 265-273.

[17] Treacy M. M. J. (1999) Dynamical diffraction in metallic optical gratings. Appl. Phys. Lett. 75: 606-608.

[18] Lee K. G. and Park Q. H. (2005) Coupling of Surface Plasmon Polaritons and Light in Metallic Nanoslits. Phys. Rev. Lett. 95: 103902.

[19] Takakura Y. (2001) Optical resonance in a narrow slit in a thick metallic screen. Phys. Rev. Lett. 86: 5601-5603.

[20] Yang F. and Sambles J. R. (2002) Resonant Transmission of Microwaves through a Narrow Metallic Slit. Phys. Rev. Lett. 89: 063901.

[21] Martin-Moreno L., Garcia-Vidal F. J., Lezec H. J., Pellerin K. M., Thio T., Pendry J. B., and Ebbesen T. W. (2001) Theory of extraordinary optical transmission through subwavelength hole arrays. Phys. Rev. Lett. 86: 1114-1117.

[22] Salomon L., Grillot F., Zayats A. V., and Fornel F. (2001) Near-field distribution of optical transmission of periodic subwavelength holes in a metal film. Phys. Rev. Lett. 86: 1110-1113.

[23] Martin-Moreno L. and Garcia-Vidal F. J. (2004) Optical transmission through circular hole arrays in optically thick metal films. Opt. Express 12: 3619-3628.

[24] Hohng S. C., Yoon Y. C., Kim D. S., Malyarchuk V., Muller R., Lienau C., Park J. W., Yoo K. H., Kim J., Ryu H. Y., and Park Q. H. (2002) Light emission from the shadows: Surface plasmon nano-optics at near and far fields. Appl. Phys. Lett. 81: 3239-3242.

[25] Kim D. S., Hohng S. C., Malyarchuk V., Yoon Y. C., Ahn Y. H., Yee K. J., Park J. W., Kim J., Park Q. H., and Lienau C. (2003) Microscopic Origin of Surface-Plasmon Radiation in Plasmonic Band-Gap Nanostructures. Phys. Rev. Lett. 91: 143901.

[26] Ropers C., Park D. J., Stibenz G., Steinmeyer G., Kim J., Kim D. S., and Lienau C. (2005) Femtosecond light transmission and subradiant damping in plasmonic crystals. Phys. Rev. Lett. 94: 113901.

[27] Gordon R. and Brolo A. G. (2005) Increased cut-off wavelength for a subwavelength hole in a real metal. Opt. Express 13: 1933-1938.

[28] Klein Koerkamp K. Enoch J., S., Segerink F. B., Van Hulst N. F., and Kuipers L. (2004) Strong influence of hole shape on extraordinary transmission through periodic arrays of subwavelength holes. Phys. Rev. Lett. 92: 183901.

[29] Degiron A. and Ebbesen T. W. (2005) The role of localized surface plasmon modes in the enhanced transmission of periodic subwavelength apertures. J. Opt. A: Pure Appl. Opt. 7: S90-S96.

[30] van der Molen K. L., Klein Koerkamp K. J., Enoch S., Segerink F. B., Van Hulst N. F., and Kuipers L. (2005) Role of shape and localized resonances in extraordinary transmission through periodic arrays of subwavelength holes: Experiment and theory. Phys. Rev. B 72: 045421.

[31] Ruan Z. and Qiu M. (2006) Ehanced transmission through periodic arrays of subwavelength holes: the role of localized waveguide resonances. Phys. Rev. Lett. 96: 233901.

[32] Lee J. W., Seo M. A., Kang D. H., Khim K. S., Jeoung S. C., and Kim D. S. (2007) Terahertz electromagnetic wave transmission through single rectangular holes and slits in thin metallic sheets. Phys. Rev. Lett. 99: 137401.

[33] Bethe H. A. (1944) Theory of diffraction by small holes. Phys. Rev. 66: 163-182.

[34] Boukamp C. J. (1954) Diffraction Theory. Rep. Prog. Phys. 17 : 35-100.

[35] García de Abajo F. J. (2002) Light transmission through a single cylindrical hole in a metallic film. Opt. Express 10: 1475-1484.

[36] Degiron A., Lezec H. J., Yamamoto N., and Ebbesen T. W. (2004) Optical transmission properties of a single subwavelength aperture in a real metal. Opt. Commun. 239: 61-66.

[37] Garcia-Vidal F. J., Moreno E., Porto J. A., and Martin-Moreno L. (2005) Transmission of light through a single rectangular hole. Phys. Rev. Lett. 95: 103901.

[38] Van Exter M. and Grischkowsky D. (1990) Optical and electronic properties of doped silicon from 0.1 to 2 THz. Appl. Phys. Lett. 56: 1694-1696.

[39] Jiang Z., Li M., and Zhang X. C. (2000) Dielectric constant measurement of thin films by differential time-domain spectroscopy. Appl. Phys. Lett. 76: 3221-3223.

[40] Kang C., Maeng I. H., Oh S. J., Son J. H., Jeon T. I., An K. H., Lim S. C., and Lee Y. H. (2005) Frequency-dependent optical constants and conductivities of hydrogenfunctionalized single-walled carbon nanotubes. Appl. Phys. Lett. 87: 041908.

[41] Han H., Park H., Cho M., and Kim J. (2002) THz pulse propagation in plastic photonic crystal fiber. Appl. Phys. Lett. 80: 2634-2636.

[42] Hara J. F., Averitt R. D., and Taylor A. J. (2004) Terahertz surface plasmon polariton coupling on metallic gratings. Opt. Express 12: 6397-6402.

[43] Torosyan G., Rau C., Pradarutti B., and Beigang R. (2004) Generation and propagation of surface plasmons in periodic metallic structures. Appl. Phys. Lett. 85: 3372-3374.

[44] Gomez Rivas J., Schotsch C., Haring Bolivar P., and Kurz H. (2003) Enhanced transmission of THz through subwavelength holes. Phys. Rev. B 68: 201306.

[45] Gomez Rivas J., Kuttge M., Haring Bolivar P., and Kurz H. (2004) Propagation of surface plasmon polaritons on semiconductor gratings. Phys. Rev. Lett. 93: 256804.

[46] Lee J. W., Seo M. A., Kim D. S., Jeoung S. C., Lienau Ch., Kang J. H. and Park Q. H. (2006) Fabry-Perot effects in THz time-domain spectroscopy of plasmonic band-gap structures. Appl. Phys. Lett. 88: 071114.

[47] Azad A. K., Zhao Y., and Zhang W. (2005) Transmission properties of terahertz pulses through an ultrathin subwavelength silicon hole array. Appl. Phys. Lett. 86: 141102.

[48] Qu D., Grischkowsky D., and Zhang W. (2004) Terahertz transmission properties of thin, subwavelength metallic hole arrays. Opt. Lett. 29: 896-898.

[49] Cao H. and Nahata A. (2004) Influence of aperture shape on the transmission properties of a periodic array of subwavelength apertures. Opt. Express 12: 3664-3672.

[50] Shin Y. M., So J. K., Jang K. H., Won J. H., Srivastava A., and Park G. S. (2007) Superradiant terahertz smith-purcell radiation from surface-plasmon excited by counter-streaming electron beams. Appl. Phys. Lett. 90: 031502.

[51] Jeon T. I., Zhang J., and Grischkowsky D. (2005) THz Sommerfeld wave on a single metal wire. Appl. Phys. Lett. 86: 161904.

[52] Jeon T. I. and Grischkowsky D. (2006) THz Zenneck surface wave propagation on a metal sheet. Appl. Phys. Lett. 88: 061113.

[53] Seo M. A., Park H. R., Koo S. M., Park D. J., Kang J. H., Suwal O. K., Choi S. S., Planken P. C. M., Park G. S., Park N. K., Park Q. H. and Kim D. S. (2009) Terahertz field enhancement by a metallic nano slit operating beyond the skin-depth limit. Nature Pnoton. 3: 152-156.

[54] Matsui T., Agrawal A., Nahata A. and Vardeny Z. V. (2007) Transmission resonances through aperiodic arrays of subwavelength apertures. Nature 446: 517-521.

[55] Williams C. R., Andrews S. R., Maier S. A. Fernández-Domínguez A. I., Martín-Moreno L., and García-Vidal F. J. (2007) Highly confined guiding of terahertz surface plasmon polaritons on structured metal surfaces. Nature Photon 2: 175-179.

[56] Pendry J. B., Martin-Moreno L., and Garcia-Vidal F. J. (2004) Mimicking Surface Plasmons with Structured Surfaces. Science 305: 847-848.

[57] Garcia-Vidal F. J., Martin-Moreno L., and Pendry J. B. (2005) Surfaces with holes in them: new plasmonic metamaterials. J. Opt. A: Pure Appl. Opt. 7: S97-S101.

[58] Hibbins A. P., Evans B. R., and Sambles J. R. (2005) Experimental verification of designer surface plasmons. Science 308: 670-672.

[59] Garica de Abajo F. J., Gomez-Santos G., Blanco L. A., Borisov A. G., and Shabanov S. V. (2005) Tunneling mechanism of light transmission through metallic films. Phys. Rev. Lett. 95: 067403.

[60] Garcia de Abajo F. J., Gomez-Medina R., and Saenz J. J. (2005) Full transmission through perfect-conductor subwavelength hole arrays. Phys. Rev. E 72: 016608.

[61] Treacy M. M. J. (2002) Dynamical diffraction explanation of the anomalous transmission of light through mettallic gratings. Phys. Rev. B 66: 195105.

[62] Garcia-Vidal F. J. and Martin-Moreno L. (2002) Transmission and focusing of light in one-dimensional periodically nanostructured metals. Phys. Rev. B 66: 155412.

[63] Kang J. H., Park Q. H., Lee J. W., Seo M. A., and Kim D. S. (2006) Perfect transmission of THz waves in structured metals. J. Korean Phys. Soc. 49: 881-884.

[64] Lee J. W., Seo M. A., Park D. J., Jeoung S. C., Park Q. H., Lienau Ch., and Kim D. S. (2006) Terahertz transparency at Fabry-Perot resonances of periodic slit arrays in a metal plate: experiment and theory. Opt. Express 14: 12637-12643.

[65] Tanaka M., Miyamaru F., Hangyo M., Tanaka T., Akazawa M., and Sano E. (2005) Effect of a thin dielectric layer on terahertz transmission characteristics for metal hole arrays. Opt. Lett. 30: 1210-1212.

[66] Lee J W., Seo M. A., Park D. J., Kim D. S., Jeoung S. C., Lienau Ch., Park Q-Han, and Planken P. C. M. (2006) Shape resonance omni-directional terahertz filters with near-unity transmittance. Opt. Express 14: 1253-1259.

[67] Lee J. W., Seo M. A., Sohn J. Y., Ahn Y. H., Kim D. S., Jeoung S. C., Lienau Ch., and Park Q-Han (2005) Invisible plasmonic metamaterials through impedance matching to vacuum. Opt. Express 13: 10681-10687.

[68] Zhao G., Schouten R. N., Van der Valk N., Wenckebach W. Th., and Planken P. C. M. (2002) Design and performance of a THz emission and detection setup based on a semi-insulating GaAs emitter. Rev. Sci. Instrum. 73: 1715-1719.

[69] Saleh B. E. A. and Teich M. C. (1991) Fundamentals of Photonics. John Wiley & Sons, Inc.

[70] Wu Q. and Zhang X. C. (1995) Free-space electro-optic sampling of terahertz beams. Appl. Phys. Lett. 67: 3523-3525.

[71] Lee J. W. and Kim D. S. (2009) Relative contribution of geometric shape and periodicity to resonant terahertz transmission. J. Appl. Phys. 107: 113109.

[72] Nau D., Schönhardt A., Bauer Ch., Christ A. I., Zentgraf T., Kuhl J., Klein M. W., and Giessen H. (2007) Correlation effects in disordered metallic photonic crystal slabs. Phys. Rev. Lett. 98: 133902.

Permissions

The contributors of this book come from diverse backgrounds, making this book a truly international effort. This book will bring forth new frontiers with its revolutionizing research information and detailed analysis of the nascent developments around the world.

We would like to thank Ki Young Kim, for lending his expertise to make the book truly unique. He has played a crucial role in the development of this book. Without his invaluable contribution this book wouldn't have been possible. He has made vital efforts to compile up to date information on the varied aspects of this subject to make this book a valuable addition to the collection of many professionals and students.

This book was conceptualized with the vision of imparting up-to-date information and advanced data in this field. To ensure the same, a matchless editorial board was set up. Every individual on the board went through rigorous rounds of assessment to prove their worth. After which they invested a large part of their time researching and compiling the most relevant data for our readers. Conferences and sessions were held from time to time between the editorial board and the contributing authors to present the data in the most comprehensible form. The editorial team has worked tirelessly to provide valuable and valid information to help people across the globe.

Every chapter published in this book has been scrutinized by our experts. Their significance has been extensively debated. The topics covered herein carry significant findings which will fuel the growth of the discipline. They may even be implemented as practical applications or may be referred to as a beginning point for another development. Chapters in this book were first published by InTech; hereby published with permission under the Creative Commons Attribution License or equivalent.

The editorial board has been involved in producing this book since its inception. They have spent rigorous hours researching and exploring the diverse topics which have resulted in the successful publishing of this book. They have passed on their knowledge of decades through this book. To expedite this challenging task, the publisher supported the team at every step. A small team of assistant editors was also appointed to further simplify the editing procedure and attain best results for the readers.

Our editorial team has been hand-picked from every corner of the world. Their multi-ethnicity adds dynamic inputs to the discussions which result in innovative outcomes. These outcomes are then further discussed with the researchers and contributors who give their valuable feedback and opinion regarding the same. The feedback is then collaborated with the researches and they are edited in a comprehensive manner to aid the understanding of the subject.

Apart from the editorial board, the designing team has also invested a significant amount of their time in understanding the subject and creating the most relevant covers. They scrutinized every image to scout for the most suitable representation of the subject and create an appropriate cover for the book.

The publishing team has been involved in this book since its early stages. They were actively engaged in every process, be it collecting the data, connecting with the contributors or procuring relevant information. The team has been an ardent support to the editorial, designing and production team. Their endless efforts to recruit the best for this project, has resulted in the accomplishment of this book. They are a veteran in the field of academics and their pool of knowledge is as vast as their experience in printing. Their expertise and guidance has proved useful at every step. Their uncompromising quality standards have made this book an exceptional effort. Their encouragement from time to time has been an inspiration for everyone.

The publisher and the editorial board hope that this book will prove to be a valuable piece of knowledge for researchers, students, practitioners and scholars across the globe.

List of Contributors

Guy A. E. Vandenbosch
Katholieke Universiteit Leuven, Belgium

Paolo Di Sia
Free University of Bozen-Bolzano, Bruneck-Brunico (BK), Italy

Xingyu Gao
Institute of Opto-mechatronics, Guangxi Key Laboratory of Manufacturing System & Advanced Manufacturing Technology, School of Mechanical & Electrical Engineering, Guilin University of Electronic Technology, Guilin Guangxi, China

Brian Ashall
School of Science, Technology, Engineering and Mathematics, Institute of Technology Tralee, Ireland

Dominic Zerulla
School of Physics, College of Science, University College Dublin, Belfield, Dublin 4, Ireland

Yongqi Fu
School of Physical Electronics, University of Electronic Science and Technology of China, China

Jun Wang and Daohua Zhang
School of Electronic and Electrical Engineering, Nanyang Technological University, Republic of Singapore

Valeria Lotito
Electronics/Metrology Laboratory, Empa, Swiss Federal Laboratories for Materials Science and Technology, Dübendorf, Switzerland
Department of Information Technology and Electrical Engineering, ETH Zurich, Zurich, Switzerland

Christian Hafner
Department of Information Technology and Electrical Engineering, ETH Zurich, Zurich, Switzerland

Urs Sennhauser
Electronics/Metrology Laboratory, Empa, Swiss Federal Laboratories for Materials Science and Technology, Dübendorf, Switzerland

Gian-Luca Bona
Direction, Empa, Swiss Federal Laboratories for Materials Science and Technology, Dübendorf, Switzerland
Department of Information Technology and Electrical Engineering, ETH Zurich, Zurich, Switzerland

Baltar Henrique T. M. C. M., Drozdowicz-Tomsia Krystyna and Goldys Ewa M.
Faculty of Science of Macquarie University, Department of Physics and Astronomy, Sydney, Australia

V. A. G. Rivera, F. A. Ferri, O. B. Silva, F. W. A. Sobreira and E. Marega Jr.
Instituto de Física de São Carlos, Universidade de São Paulo, São Carlos, Brazil

Joong Wook Lee
Advanced Photonics Research Institute, GIST, Gwangju, Republic of Korea

DaiSik Kim
Department of Physics and Astronomy, Seoul National University, Seoul, Republic of Korea